SCIENTIFIC EXPLANATIONS

austintorn@aol.com

BRAIN WAVES

< 1 >
— Referred Touch —

Take a pencil, and feel a texture with it—
You seem to feel it at the pencil's tip;
Yet, you have no sense organs way out there;
So—the brain fabricates reality!

< 2 >
— Reality is Real-ized —

All we 'see' are the insides of our heads,
A model of reality. No, you say—
Well, it's the same model 'seen' in your dreams,
With your eyes closed and you in darkness, too!

< 3 >
— A Useful Illusion —

Not only is 'seeing' inside our heads,
But also the hard-soft-texture of touch,
The scents of molecule shapes, and the sounds
Of air waves, again, as in a night dream.

< 4 >
— Tuning In —

Reality plays within our theater,
A wide-awake dream. What's really out there,
I suppose, are waves and fields, which our minds
Sense in representative ways, like 'red'.

REALITY UNVEILED

NIL

Something from nothing won't do;
Nor something from infinite causes, too;
Thus, forever never began.

The Prelude

It has been raining for a few days now
In the Appalachian mountains of New York,
The very rains that flooded Georgia.

They had a long drought there,
And prayed for its end,
Led by their holy Governor,
But the drought continued.

Perhaps now they pray for the rains to stop,
But it's still a very rainy night in Georgia.

As we have been fine-tuned by evolution
To exist in our little corner of the universe,
The water falling from the sky
In its pattern of circulation from evaporation
Is in our best interest, in the general sense,
That is, but not so much in the specific sense
Since any one region may get too much or too little;
The farmer has different wishes than those
In the wedding party across the road.

My office is usually outside on the back deck
For the three fair seasons,
Even under a big umbrella in light rain,
But, due to the heavy downpours today,
I must soon retire to my alternate office
The spacious boiler room
That's partway underground.

It looks like the end of the world outside now,
It being almost dark at 3 PM,
A fitting day to write about The Beginning,
So to speak, although a beginning there never was,
For 'something' was always around and about, obviously,
For the lack of 'something' is not in evidence here,
Nor could "it" have been productive, as it as no being.

While that is elementary,
It is not the total unveiling of reality,
But is merely the 'why'
Of the natural state of affairs.

And, while the causeless not being able
To have any intent or definiteness to it
Is indeed the ultimate SuperToe,
We must still explain 'how' it operates.

I keep the boiler room dark,
Even though it has a light,
The better for the screen of the laptop
To shine in all its colorful glory.

So, I traverse the darkness now, like a ninja,
Led by the glow of the computer's
Pulsing sleep indicator,
Knowing that nothing is in the way.

Here, in this semisecret chamber,
Is the one and only jewel-encrusted edition
Of the 'Great Omar' that I fished up from the Titanic
Lying on the floor of the North Atlantic.

The Rubáiyát Publisher's Gem

These pearls of thought in Persian gulfs were bred,
Each softly lucent as a rounded moon;
The diver Omar picked them from their bed,
Fitzgerald strung them on an English thread.

The Rare Book

The lone jewel encrusted 'Great Omar',
Now worth over 20 million dollars.
Sunk, with the mighty Titanic.
I plucked it up from the North Atlantic.

Here, too, the cane of GrandMaster West,
Whose adventures are detailed
In 'Butterflies At the Edge of Forever'.

Toe Questors from www.toequest.com
Discover the Secrets of the Universe,
As well as the humorously dangerous implications
That follow their possession of the Holy Grail
Of the genuine Theory of Everything.

With the world's future hanging in the balance,
They sharpen their wits and skills
Through the teachings of the learnéd Grand Masters.

Here, as well, Aristotle's 'lost' book,
'Beyond Metaphysics', one of its many variants
Obtained from the Library of Babel—
A repository that contains all possible books,
Most of them unintelligible.

"Anything?"

"No, mostly gibberish,
But I found one on a table
That someone must have treasured."

"Oh, yes, he spent his entire lifetime here.
It's Plato's 'Beyond Metaphysics'."

"Wow! That's been lost for thousands of years.
But is it the true version?"

"Who knows."

"This library contains no information whatsoever!"

"True, but there's another library next door
That also claims to have Everything."

"You mean that little 'hut'—no, wait—I get it
The library next door is empty."

"Yes, the All sums to the None."

Here, too, I have some nuggets of gold
Found in the original Garden of Eden
That was located in the heart
Of the Amazon Jungle,
Wherein lie massive fields of Lady's Slippers
And all of the flowers of Paradise.

I reached up—and put the apple back on the tree.

And the Celtic Chronicles, I have, as well,
That I found in an iron box
Beneath Glastonbury Abbey,
And, from the tomb of the Holy Sepulcher—
The Grail itself, as told in 'Last Knight's Almanac'.

The best definition of the word "almanac(k)"
Is the word itself, that is,
ALL MAN's KNACK,
Or sometimes, ALL MY KNACK.

Too, the secrets of the DIA,
All of which "never happened",
And the 'Astronomical Wonders of Outer Space',
'The Triumph of Life, Love, and Being',
'The Magical Moments of Life',
And 'The Universal Day',
Along with the original drawing
Of Fredrick's separation of the forces
That shows the weak and the strong forces
In their opposition and the electrical
And the magnetic in their transition.

[electric<—>magnetic] & [weak vs. strong]

Here, as well, a sliver of the True Cross,
A small vial containing a drop
Of the Virgin Mary's milk,
A pebble, from a moon rock,
Given to me by a polymath
Who works for the President,
A smart thinking
And talking cricket named Crick,
The tip of the spear that pierced
The side of the Saviour,
A few molecules of immortal air from Egypt,
Some secret papers retrieved from the shaft
Of the bottomless CIA trash pit,
A thriving rose bush, just outside the window,
That was begun from Khayyàm's
11th century garden,
'Flamberge'—Prince Valiant's 'Singing Sword'
(Twin to 'Excalibur'),
Thomas Jefferson's briefcase,
An original and intact Ming Dynasty vase,
The third [missing] tablet of Commandments,
And, on the wall, hanging, Nobody's epitaph,
Written by himself and now emblazoned on his tombstone:

Hitherto whereby whence we came?

Are we talking consciousness relative to 'C'
Or fully present as events they occur
Thence they went?

Do we see a dead star shining brightly
In the fully present Now
Albeit partially past then?

Here, too, one of ProfPat's protons
That can never roll off the end of the desk,
Ever stopping at the edge, and returning,
As well as an enlarged copy of the WMAP survey—
A blank spot in it perhaps indicating
A universal size collision.
And, too, the solution to gravity,
And it as means for the quantum collapse,

As well as a tennis ball with my initials
Marked on it in a yin-yang style.

Yet, all of these treasure pale in comparison
To Reality's Truth unveiled.

And so this Truth shall be revealed, presented,
Displayed, shown, exhibited, released and launched,
Brought out and disclosed, divulged and made known,
Broadcast and communicated, in due time,
But, one must prepare to receive it.

In it, science and religion converge, but not quite;
Everything meets Nothing, almost;
The definite and the indefinite approach,
But just nearly; however,
Beauty does become the meaning of Truth.

All of these treasures pale in comparison
To Reality's Truth unveiled.

Or do they?
We will see if they do in the part
Entitled 'Essence and Existence'.

Reality Unbound

Now, even if 'reality' is a dream or a hologram
Put in place by a Great Projector,
Then this 'projector' is still a real 'something',
Not to mention the mechanism of
Letting its dream subjects feel the world
And its sensations as real, being real,
Along with the simulation of every subatomic interaction
Such that all is functional and not just for show,
As in a night dream.

This is where faith (making something up)
Stops way too soon, for projectors have parts.
This shows that the projector
Is but an intermediate stage, and, if so, fine,

And then we can go on past this regress
And get back to the business of determining
The first, and uncaused, fundamental.

One must fully explain the ramifications
Of a claim or a word or else it is just a magic show.
Just saying that a God or a projector
"Did it" goes nowhere at all.

It ends nothing and says nothing.
The only 'meaning' of the use of magic words
With no foundation is "Halt; that's it;
Done; mystery over; question not answered,
But only made into a much larger question".

This is not to say that we can't have theories,
But presenting theories as fact,
Instead of saying that it is just a theory,
Especially of invisible imaginary concepts,
Is just plain deceptive, a practice made even worse
By presenting it to the young minds of children as truth.

But, the believer wants to believe the myth.
So what! One's wants don't equate to the truth.

So, then, what about that First 'something',
It necessarily having been around for all time—
The causeless eternal ground-state?

If there had been nothing to it
Then there would still be nothing 'here'.

So, it was natural and normal, obviously,
For there to be a 'something',
Since it is here and does something.

Thus, we have the simple Why, but, as always,
We wish to delve into the very nature
Of the basis to learn the How.

In my office, I have the 'treasure'
Of a preliminary but solid indication

Of the Higgs particle's existence,
Which Lisa Randall was nice enough
To give me from the LHC's latest analysis.

It is but a piece of the puzzle of what already is,
So, fine, but "what is" is only an effect
Of the nature of reality's bloom.

We wish to know the basics—
And certainly one beyond the unwarranted leap
Of assumption that God, or a projector,
"Did it" case-closed kind that's an
Invented pronouncement fabricated in the mind.

Necessarily, we can't get at the 'something' directly,
Unless it is still available somewhere about us,
As it must be, for it is eternal,
Such as in the quantum realm,
But that kind of really direct science is yet to come,
Although very close, already, and so we can
Also use our reasoning powers.

We could even stop here, knowing for sure
That the nonsense of those proclaiming fantasy
As gospel is unethical and imaginary,
Not to mention that they
Lack any and all proof
Of the extraordinary "sure thing".

Proclaiming life to be inexplicable
But for another Life beneath it as a cause
Only serves to say that there
Had to be a LIFE behind that, and so forth...
Atlas standing upon turtles... all the way down...
From which the effect could never reach us,
Having to progress through infinity.

We could even put all those
With contradictory variations of
The ungrounded pronouncers in a room
And let them babble and just say "this is so".

At least they might note the arbitrariness
Of the other "sure" declarations that eve
They might have made had their social,
Geographical or familial origins been different.

Why people believe in weird and unknown things
Is a whole 'nother subject
Best addressed by Michael Shermer.

So, then, a magic God Projector did everything
Through infinity for all eternity
And we will burn if we don't believe it.
Done, finished. Let's go preach it
And convert everyone. Just kidding.

So, there's no getting away from
The original state having to be 'something',
Rather than Nothing or an Infinite Regress,
And being causeless at that, and eternal, to boot.

So, what if its mechanics were, say,
To generate a hologram.
It's what results that counts,
Not the means and implementation by which it does it

So, again, what does the mere messenger matter
Over the actual message itself?

This is true even if there is no real message content
Of any actual information, meaning that it
Could very well be that there is no cause or purpose
Of the original 'something' since
It never could have been given,
The demands and wishes of humans not withstanding,
Not making it be so one bit by their desires
For purpose, rewards and whatnot.
The mechanics are always interesting,
But, they are not the TOE.

Particular events and specific places,
Such as those in fiction novels,
Or those of our 'reality' are ever

Entertaining and interesting,
But, it is the universals beneath that we are after.

There could be even more than one universal,
Indicating some kind of 'separation'
Within the supposed unity.

I am always one after the universals, then,
And many of my poems and stories
Have universal themes overriding
The particular specifics, as well.

Again, we could stop, but there's little else to do
At certain times in our intellectual lives...
So, we ponder, as we love to,
At least admitting to our curious nature of inquiry,
Some of us exalting in the mystery
Rather than shunning it with "the end".

It's not guaranteed that we will get anywhere,
But that would be information in itself
Showing that it matters not if we can't,
For who could be blamed for living freely?

Finding the simple causeless
Would indeed be the [SuperTOE itself,
As there wouldn't be any more to it.

Thus, either way, we are as free to be as a bee
Within its form to smell the roses and enjoy the nectar.

"Must do this; must do that to be saved."
Blah, blah, blah.
Just arbitrary contradictions clashing away.

What, then is Everything?
One one hand it is the total of all that always was
And what it generated or combined into,
But, on the other, it is how the original stuff operates,
Its Why and How and all that has become of it,
Such as the Where, the What, the Then, the Now,
The When, and the Who, which is us.

Reality must ever be of the real, at its base,
No matter what secondary forms and particles
It comes up with and then emits to immerse.

Even the so-called Brahman would have to be real.
And what is of the real must be real as well,
And, although durable, is not necessarily ever-during.

Essence and Existence

Lest we forget the place of everyday life
Compared to that of the Quest For Everything,
Let us place it in its proper and totally superior position.

It was the above thought of precedence [in importance]
That I pondered as I ran my fingers along the blade
Of Prince Valiant's 'Flamberge',
Although not upon its edge,
For that would have drawn blood
Without even the slightest press against.

This sword was forged from metal
That was not of the Earth,
But from that of a meteorite.

Which is more important?
Existence or its essence?

Well, existence is what we have been thrust into,
Like it or not, unasked, and so it <u>must</u> be dealt with,
While essence need not be, that much,
Nor perhaps could it ever be known
In all its details that reside
So many orders of magnitude below
Within some number of nested Chinese boxes
[Yet, we will fly past all those].

What can a proton say?
It's markings are neither holy nor unholy.
It is mere stuff, made of the quarking emanations.

What really gets interesting is the complexity
That emerges from the simple's combinations, such as life.

Besides, we know existence,
First hand, for we are it!

However, our state of being doesn't even open a window
Into the next level down of the states beneath,
Much less to the ultimate essence
That is even beyond and beneath what glory shines
In the heart of a glowing quark.

We don't directly feel atoms moving,
Electric pulses traveling,
Or our bonding hormones operating.
Thus it is that introspection alone
Will never suffice, for science is required.

So, anyway, existence has the highest priority,
It trumping essence almost out of the picture
But for some quiet reflective moments
Of "What's it all about?"

Of course, the discovering and learning
Of the facts of current and prior existence
Such as those revealed through the natural sciences
Of biology, chemistry, physics, neurology,
Evolution, genetics, and so on
Will ever inform us on our existence
As it has come to be,
Being a real road that leads in the direction
Of the actual home of essence
Rather than to some ghost town.

Knowing that doth allow existence become even better.

Existence, or even its "ego"
Is not a 'wrong' message to be thrown away,
Ignored, pretended not to be real or useful,
Or to be constantly waved aside or wasted,
Unless one wishes to dump this life
And the universe into the trash.

Existence is exactly what we are.

Blah, blah blah...
Extinguish the self that nature made...
Don't use the brain fashioned over millions of years—
One so expensively wrought by evolution
And ever needing of a large part of our energy,
Oxygen, and blood...

Blah, blah, blah. All is illusion, blah, blah.
Consult the invisible and undetectable soul...
Blah, blah, blah.

The priest slowly waves the smoke-filled censor
Down the aisle to dispel the evil spirits...
This is bleak, dull, and imaginary.

Life is to be lived.
Not to do so only ever moves one toward death.
Everyone dies, but not everyone lives.

Those not busy living are busy dying.
Life is here.

What is Life?
One must live it fully to know the answer.
Technically, life is metabolism,
That is, it takes stuff in and utilizes it.

So, existence 'precedes' essence,
Even though essence came first.

We <u>are</u> the Cosmos come to life.

So, then, who cares about
'The Theory of Everything'
But as an intellectual exercise!

Live, love, and dance to enjoy
The triumph of life, love, and being.

I seem to be reversing myself here.

Is finding the deepest nature of reality
Important to us or not?

Yes and no.
All of life goes on quite well without it;
Yet, some of life is warped by
The presentation of imagination as fact.

It is not enough that no evidence exists, at all,
Much less any extraordinary evidence
For the extraordinary claims preached as truth
To young children through merciless indoctrination.

Myths will not get the world through its crises,
Some of them even caused by those
Yet mired in the mythic age.

The Box of Truth opens by itself,
No matter any delays from those trying to prevent it.

The search for the narrowing of the Truth of All
Can take us beyond the narrow confines
Of the made-up legends,
Although it may ever evade us, at the worst,
But at least it becomes better cornered, or even found,
At its best, as we will do here.

This quest, too, can inform our existence,
To some extent, as much as it may, always and ever;
But life will still go on, day-to-day, as it must,
The finds not really so much mattering to those
Who already embrace the human condition
And its joys and sorrows,
Those aspects being the necessary
And dual sides of Beauty enshrined
As both the dark valleys
And the mountainous heights.

Again, we could stop now, much remaining the same.
There is love, sunshine, friends, nature,
Adventure, fun, good sleep and nourishment.
What more could a human being want, really?

Some, of the never so humble,
Will always want more,
Such as the reward of an after life.

Anything and Everything

I am now holding part of a brick
That came from Nero's very recently discovered
Revolving banquet hall that kept pace
With the turn of the Earth.

I am about to ponder the existence of this brick,
But that would probably be too disruptive to my life,
So I'm going out to date some old fossil instead.
...
I'm back—and she is very young at heart
And quite exciting, so we are trying to tone it down
By smoking some pot and pondering the brick.
Just kidding.

Actually, I'm thinking of the Library of Congress,
For I heard that it has five hundred miles of stacks.

It began anew, after burning by the British,
When Thomas Jefferson donated his personal library.

I found his personal diary in the lining of his case.
It said the Founding Fathers wanted
To retain a Deity to save the new nation
From the religious superstitions associated with a Theity.

What does this have to do with Everything?
Well, nothing.
What does Nothing have to do with Everything?
Just about everything.

The Library of Congress is a relative nothing
Compared to the Library of Babel,
Which, in turn, is but a mote of dust
Compared to all the interactions
And arrangements of Everything;

Yet, the sum total of all possible states is nothing!

How could this be?
Well, I am speaking of the sum
Of the information content of every possible book
Or every possible arrangement as summing to zero
Since any given one is arbitrary,
Being neither wrong nor right,
But just a possible path,
Although some remain coherent longer,
Which I have to admit is at least
A degree of more workingness,
If not more rightness.

We are on one of those particular paths, and so,
While it was arbitrary, there is a least
Cause and effect that follows thereafter.

Even in our own working universe,
The positive energy of matter matches
The negative energy of gravity, all summing to zero,
Or at least as close as it can get,
For an absolute and total zero is not conceivable—
In any meaning of the word.

As we probe the 'fundamentals' composing our reality,
We note that they are extremely tiny, really,
Compared to the more massive
And composite complexities made from them.

A 'small' split-end of hair may contain a billion atoms,
A little bit of water a trillion or more.

Even an electron is as small and remote as second base
Within the relative ballpark of an atom.

As for quarks, no one has seen them directly
But for the circumstantial evidence
Of their energy jets emerging.

Equations define the gluon/fields so well
That we don't even need to experiment with them

Or see them to know their ways,
For the math already tells us.

What gives rise to these tiny 'particles'
Is smaller yet—the quantum fluctuation.

The swarms of particles emerging,
Both the enduring and the virtual
(a misnomer since they are still real)
Form the 'ether', a word that Frank Wilczek
Prefers to replace with the term 'grid' in his book,
'The Lightness of Being'.

It is also that these fluctuations
Are but appearances from a basic fundamental medium
That then waver all around, but no matter,
For, in either case we still have a causeless ground-state,
And thus it this ground-state that is eternal,
Having never been created [thus, no Creator]
And always having been, as we see,
For energy can neither be created nor destroyed.

Things get smaller, ever smaller,
Reaching a limit at the Planck size.

Strings?
Who knows, but the universal trend
Is ever towards the simpler to the simplest,
This Final Place not being where we would expect
To find the ultimate complexity, or, for that matter,
Even any specific order or anything specific,
Since there was nothing prior to define it.

Here, I am even skipping over the formation of universe
To get at the state beneath, one that is the First,
For Everything arose from it—
Universes and more quantum realms alike,
Just as all humans and animals
Evolved from a common ancestor.

It seems that the First,
Something akin to the quantum fluctuation,

Would be the simplest near 'nothing' state possible,
One about as close to Nothing as it could be,
Since the inward direction of All has a limit,
At least in our universe, making the size scale absolute,
Not relative, while the outward reach is quite large,
Perhaps going up to the mass of
What becomes a black hole,
Or even beyond that mass or size
For any stable structures that are spread out enough
Not to collapse, while also not dispersing.

Does the realm of the First, causeless, something
Lie beyond our 4-dimensional universe
And its vacuum energy,
Such as a 5th dimensional state
In which every arrangement possible
Exists in a kind of superposition,
An indefinite chaos of All,
With some kind of potential to it to generate
The specific and definite 'particles'?

This is admittedly extra fanciful,
Although a logical extension of dimension.

Doesn't matter much, although fun to talk about,
Since there is always be the sure notion
Of the causeless, eternal ground-state.

Or. alternately, was there simply some amount
Of eternally durable energy or substance
Lying about for all time?

It would have been a finite amount,
For we are not packed in like sardines;
But what defined its size, nature, and amount?

Well, that is a problem for something so discrete
And definite that existed for all time,
And so I'll concentrate more
On a kind of opposite pair emission model,
For then there can be any amount
[fluctuating, but ever summing to "zero",

As gravity cancels mass-energy],
Anywhere, any time, any amount;
There, of course, also, necessarily having to be
No special where, time, or amount for any happenings.

That aside, back to alternatively, what could energy do,
If it was always around, to make something more?

It could flow in any direction,
And, so, two opposite currents could pass each other,
Resulting in the only event possible,
That of swirling about each other,
Eventually perhaps winding into something larger
Until it could be compressed no more,
Then, continuing to throw off the excess
At its spinning speed of light, perhaps,
Then build into higher whirling structures.

It might take millions of years
To form even one atom,
But... This is Jimbo's FET theory
(Fluid Energy Theory).

Again, alternately, what could substances do?
It could run into each other and interact,
And, to simplify, build to more complex structures.

The problems yet remain
As to what determined the size,
Nature, and the amount of the
First and Original energy or substance,
And, so, we might as well ever go back
To the causeless chaos of the indefinite,
Which, like anything that is the First,
Could not have been previously defined,
For there was no 'previous' state.

Anyway, the veil of reality lifts.
Take your pick, for there is ever and still
The eternal causeless ground-state.

So, was our universe a matter
Of luck and happenstance,
A chance that was made more likely,
Given all of eternity?

Well, it does seem that any design
Could not itself be absolute and fundamental.

Wait long enough, and then even the rarest of probabilities
Happens from all the possibility, such as our universe.

The key point, but not an explanation of operation,
Is that the initial, eternally existing 'something'
Would have no particular order,
No pre-thought design, no 'smart' system aspect
Apart from the natural aspects of the 'something',
For, again, there was nothing prior to intend it.

It is our science that attempts to understand
The workings of the path our universe has taken,
Ever explaining away the mysteries
As it marches on and on.

We are, though, now, quite beyond and beneath
The dimensions, laws, and forms,
Where it is all generated, this being a lawless,
Formless, placeless entity—an "undigested mass".

This is where religion diverges from science,
Both having converged upon the causeless
For a perfect and definite God-order
Is the opposite of an indefinite disorder.

And, too, Everything almost converges to Nothing,
But for the tiny simple 'something'
Of the potential of the quantum fluctuations,
Otherwise called tunneling or uncertainty.

What could generate the secondary 'stuff',
Without a plan, if it did?

Well, something that took all paths, by brute force,
Such as all the possible ones happening
In kind of a 5th dimension of superposition
Like that we know of in the quantum realm,
Spitting out some particles that work or don't.

Again, this mechanism is not so important in itself, Although interesting, for, the ground-state ever was.

So far we have traveled the unlit
Corridors back toward Antiquity.

There's no dust and cobwebs,
For they are of definite, although,
As usual, secondary, forms.
There is only darkness (no photons).
...
We have begun to build quantum computers
That will be eventually be able follow
All the paths of a computer program at once,
Rendering obsolete, for example,
Encryption codes involving large prime factors
Of very large prime numbers;
But they are in their infancy,
Having at most a handful of qu-bits
And also requiring massive shielding,
Lest the wave functions decohere
Due to the noise or gravity of everyday existence.

However, the First, natural,
Quantum-type superposition of All possible paths would
Have needed no shielding,
For it was all there was.

How long would it take for the near infinite amount
Of paths to be traversed,
Even going down them "all at once"?

So, did it, by leaving no path unfollowed,
Happen to produce some stable and enduring particles,
Among numerous failures that soon went back
In, or noplace really fast, remaining inert)?

Maybe, but what observation or interactions
Collapsed the wave function from the outside
When there were no observers,
If we even believe this requirement, and no outside?

And, since there was no collapse possible, did it,
In its superpositional potential form,
Continue to go down the paths of the possible particles
Combining and evolving into stars spewing elements
And all that right on up through cells happening
And what we know as evolution improving the species,
All of this occurring, as said,
While still operating in the "everything-possible" mode,
Continuing on right up to the point
Where animal or human consciousness appeared—
To finally collapse the universal wave function
From the inside, actual-izing and real-izing
Everything by conscious observation?
Whew! Well, that was fun,
But, again, it's not required.

Or, is it simply that, given all of time,
That some particles with the right stuff
Would eventually appear and form a working universe,
The rest a colossal waste if not recycled somehow?

Or, stranger yet,
Are we still within the wave function,
As Stephen Hawking suggests,
Our universe spending the majority of its time
In its most probable working state?

Again, take your pick,
But at least we have honed in
On the deep neighborhood of reality's origin,
Lifting up a corner of the veil of the deep dark secret
That could really jolt humanity,
Turning the world inside out,
For, even now, they must ever face
this eternal causeless ground-state
Of there being no intent or direction to it.

OF CAUSE THE FIRST

Whatever is eternal and is so well defined
Could never be as so, for it was never defined
In the first place, for that there never was
To define all that it forever did and does.

I hold in my hand a bone
From Early Sapiens or of Proto-man.

He is not gone, though,
But lives on in your heart and mine,
As in him lived all those before
In which the universe itself came to life.

Someday we'll evolve into a Universal Mind,
As the Ultimate Complexity is yet to come,
For it is to be found beyond, in the future,
Not beneath, in the past.

We were looking in the complete wrong direction.
Ever look and push forward,
Rather than stumble backwards.

NAPOLEON

I wrote to the banished Napoleon,
Asking if all was well, and he replied,

No, sir.
PS: Liveable was I ere I saw Elba, evil's prison.

(I extended the palindrome;
However, there is one cheat...)

THE AGE OF JUNE

Now June embodies us—
It is the hinge of season and of life,
So—take heed, fond man,
And pass some few years
As the full blossom of the June rose,
For these are rare times now.

Soon enough comes the autumn of care
Sobering into age, thence into
The pale white winter of death,
And not yet the warm indolent summer
Of contentment lazing into mid-age,
Though surely past is the crisp,
Flowering youth-spring of joy!

Behold now thy pictur'd life in June,
A nameless, happy season well spent
Between passion and contentment,
A time when life is made or not;
For I am now June,
And June is me in age,
And I can stay, or go either way.

SULTRY NIGHTS

I awoke, her scent traveling,
All my senses merging, changing,
My hearing following the one vibration
That echoed from below—
A spirit, leading me to the lake,
Plunging me into the cooling depths,
Where the mermaid waited for me.

THE IRRESISTIBLE MEETS THE IMMOVABLE

She wanted to charge all those irresistible objects,
Even straight out of her husband's pockets;
But that grouchy and unlovable force
Said she could but window shop, of course.

So she bought stain-glass windows,
All very expensive ones, too,
For all the bedrooms and the baths,
A very large number, in her wrath.

She went to the hospital for the emergency
Of the credit card cut of plastic surgery;
'Twas nothing they could do for the charge,
Unable even to puzzle it together at large.

She returned home very much negative,
Finding him accounting all his positives.

"Stick 'em up," said she, assertively,
A gun pointed at his fatherlies;
"Give me your money or your life
If you want me as your loving wife!"

Said he, "Do I need all this rife and strife?
Go take up music and blow a fife!"

"When I was making the lion's fare,
All of that with you I did share.
I need to glamorously amorous
To keep our relationship harmonious."

He opened a drawer in his brain, that Mister,
That one of his mental cash register;
He then closed this manly box, in fact,
And into his brain he put it back.

So, on and on this all ever went,
But it was still that nothing could be spent.
She robbed his wallet while he slept
And left all the house very unkept.

Paternity and Maternity soon had it out
This battle an evenly matched bout;

The irresistible force met the immovable object,
Of which result no one could ever suspect.

What would happen, nobody knew;
And then everything really blew;

A total annihilation had occurred,
Nothing left at all but for a blur
A noisy bang, and a flash of light,
Such as with all creation's might.

All had been shredded and torn,
As when the universe was born.

(Brain Waves Continues, wherever)

< 5 >
— A Receiver —

The brain is like a TV tuner, reading waves
That originate somewhere else entirely;
Hard reality is totally fabricated,
Just as in our dreams that seem so real.

< 6 >
— Making Sense —

Absolute Reality is scentless,
Colorless, and quite soundless; however,
Sense organs detect waves and vibrations—
So—all reality's fabricated.

THE PRIME MOVER

I am quite on David's side on most of this,
But neither of us have come to agreements
As to the 'prime mover' status of the universe,
And this is maybe where you come in, Bogie,
As we need the discussion of the 'prime mover'
To take place—as it's the true mystery of all forces' actions,
Which I happen to state is naturally hydrodynamic,
At wave-frequency/thermodynamic limit... — Lloyd

It could be of collisions of large arenas, as Bogie suggests,
Such as later on, but, all in all, where did theirs
And our particles come from,
As well as their amount and all that?

It would be that FS
Is Fundamental Durable Substance (FDS),
The "durable" meaning that it appears
And remains around to some extent
Rather than annihilating back.

Its prime mover would be of its source—
Of being emitted from the quantum fluctuations,
Or, one could even move the FS back to this level,
Its patterns then being taken as the quantum realm.

Either way, the base level—the ground-state,
Must be causeless, and rather arbitrary,
Due to its causelessness.

As coming from Nothing was not an option,
It had to be there, and so this all accounts
For the "prime mover".

This is so simple that it
Doesn't sound much like a TOE,
But "simple" is what the "beginnings" must be,
For higher assemblies become and increase
But later on, going onwardly upward.

So, then, the causeless IS the TOE—the prime mover.

Can't have an endless regress
Of movers behind movers.

Cause and effect is then
Only of the subsequent realm.

Nothing more would be beneath
The quantum fluctuations,
For its "possibility" needs nothing beneath it,
For this would only be more possibility anyway.

Another way to look at it is
That all the negatives and positives
Cancel back to near zero,
But for the tiny residue of the fluctuations,
And so their combined movements
Don't really count as
Being directly proportional
To the "prime movers" oomph,
Leaving the supposed
Large prime mover to be
But a rather tiny puff of
The quantum fluctuation's emanations,
Some kind of natural jiggles
That have to happen,
Else a real Nothing could appear,
Which cannot be;
For, it is the simplest possible state,
And, as such, like as that
We see in the simple states,
Ever goes into changes;
However, Nothing is so unstable
That it cannot ever be.

So, I have FQ—Fundamental Quantum,
Which leads to "FS",
Everything thereafter then being
As of Lloyd's/David's descriptive happenings.

GOD SPEAKEST TO THE POPE'S UNHOLINESS

God:
Well, Pope, of course,
Preying on the underage for sex,
Or for any reason, is WRONG,
And, as you tolerated it, even enabling it,
You allowed it to continue, even for a long time.
It is both what one does and doesn't do
That constitutes an action.
I hereby damn you
And all those involved to eternal fire.

Pope:
But we all repented by saying the Act of Contrition,
And so now all of our slates are clean.

God:
Nope, Pope, for that's just one
Of your made-up human rules,
As well as the one about
Consensual adult sex not being for recreation
But only for procreation.

Pope:
Damn; I mean darn!
If I have to burn forever, then, well, that's it;
But, are you saying that I can have sex?

God:
Of course you can;
Find some sweetheart, of nun or another;
For, your celibacy rule is complete nonsense—
Look at the Episcopals: they can marry and have kids
And all that, for that is all totally natural.
And how are your priests going to counsel families
When they don't have any familial experience of their own?

Pope:
OK, but really... is sex OK?

God:
You numbskull;
Sex is natural, too, even for the unmarried—
And stop sweating all this impure nakedness stuff.
If I didn't want people to be able to be naked
Then they would have been born with a lot more fur,
Or with clothes on.

Pope:
Anyway, should I and all resign?

God:
Yes; in fact, you are to be fired, literally;
You are all a disgrace to all that's holy.

Pope:
But, you made us!

God: Hey, well, nobody's perfect,
But, hey, don't forget: I'm the boss.

Pope:
OK, but you can't fire me, for, I quit.

God:
Good, you ignoramus.
I'm hiring a team to take over,
Consisting of Sam Harris, Michael Shermer,
Richard Dawkins, Victor Stenger, and more;
For I am an atheist Myself,
For there ain't any HIGHER POWER
That made me, that's for sure.

Pope:
How does one have sex?

God: Ah, a fine thing that question is;
You makka da rules but never playya the game.

Pope:
Are You Italian?

God:
Is the Pope decent?

Pope:
No, I am the scum of the earth.

God:
What the Hell were you thinking?

Pope:
I wasn't. So, how does sex go?

God:
It comes and then it goes. Good luck.

Pope:
Thanks; I'm going now to drink some wine,
Light some candles, and get someone out of her habit.

God:
Good plan, for that's how I,
Mister Right, found Miss Perfect.

Pope: You mean you have sex!

God:
You dumbo;
Did you think that eternity could ensue
From no paternity joining with maternity?

Pope:
Oh.

God:
Now get lost, you dingleberry,
And if you're pretty good from now on
I'll take a few trillion years off of your eternal sentence.

Pope: Thanks, I think...

SUBJECTIVE TRAPS

Many of the subjectives that are based on
Outside incoming experiences
Can indeed work well, as well as those
Intellected and emotioned from the inside,
When they all have intermixed
Of what we've actually experienced as real.

Pure introspection, though,
Along with wishes and imaginations,
May come up with invisibles,
Spirits, and realms beyond,
As these may be satisfying.

It becomes gospel because one
Is not privy, aware, or knowing
About all the hundred billion neurons
At work below,
Each with thousands of connections,
As these are the telling states
Beneath the lived state of being;
And, so, introspection, alone, cannot suffice,
For it believes that it is all there is,
That it is the one and only source of truth,
That since it thought the thought,
It must be true,
For only the enlightened few
Can disregard their own thoughts.

So, science is required,
To completely inform us all
Of all becoming known,
And use it to sort out
The mind's pitfalls.

ALL THAT LIES BETWEEN

It is a beauty and a brilliance
Flashing up in its destructance;
For, everything isn't here to stay its "best";
It's merely there to die in its sublimeness.

Like slow fires making their brands, it breeds;
Yet, ever consumes and moves on, as more it feeds,
Then spreads forth anew, this unpurposed dispersion,
An inexorable emergence with little reversion,
Ever becoming of its glorious excursions
Through the change that patient time restrains,
And feasting upon the glorious decayed remains,
In its progressive march through losses for gains.

We have oft described the causeless—
That which was always never the less,
As well as the beginnings of our quest,
And, too, have detailed, in the rarest of glimpses,
The slowing end of all of "forever's" chances;

So, then, we must now turn our attention keen
To all of the action that's here in-between—
All that's going on, and has gone before,
Out to the furthest reaches "ever-more";
For, everything that ever happens,
Including life and all our questions—
Meaning every single event ever gone on
Of both the animate and the non—
Is but from a single theme played upon.

This, then, is of the simplest analysis of all,
For it heeds mainly just one call—
That of the second law: dispersion,
The means for each and every occasion,
From the closest to the farthest range—
That which makes anything change.

These changes range from the simple,
Such as a bouncing ball resting still
Or ice melting that gives up its chill,

To the more complex, such as digestion,
Growth, death, and even reproduction.

There is excessively subtle change, as well,
Such as the formations of opinions tell
And the creation or rejections of the will.

And, yet, all these kinds of changes, of course,
Still become of one simple, common source,
Which is the underlying collapse into chaos—
The destiny of energy's unmotivated non purpose.

All that appears to us to be motive and purpose
Is in fact ultimately motiveless, without purpose;
Even aspirations and their achievement's ways
Have fed on, and come about through, the decay.

The deepest structure of change is but decay;
Although, it's not the quantity of energy's say
That causes decay, but the *quality*, for it strays.

Energy that is localized is potent to effect change,
And, in the course of causing change, it ranges,
Spreading, and becoming chaotically distributed,
Losing its *quality* but never of its quantity rid.

The key to all this, as we will see,
Is that it goes though stages wee,
And so it doesn't disperse all at once
As might one's paycheck before a month.

This harnessed decay results not only for
Civilizations, but for all the events going fore
In the world and the universe beyond,
It accounting for all discernible change,
Of all that ever gets so rearranged;
For, the *quality* of all this energy kinged
Declines, the universe unwinding, as a spring.

Chaos may temporarily recede,
Quality building up for a need,
As when cathedrals are built, or forms,

And when symphonies are performed;
But, these are but local deceits,
Born of our own conceits;
For, deeper in the world of kinds
The spring inescapably unwinds,
Driving its energy away—
As ALL is being driven by decay.

The *quality* of energy meant
Is of its dispersal's extent;
When it is totally precipitate,
It destroys; but when it's gait
Is geared through chains of events
It can produce civilization's tenants.

Ultimately, energy naturally,
Spontaneously, and chaotically
Disperses, causing change, irreversibly;

Think of a crowd of atoms jostling,
At first as a vigorous motion happening
In some corner of the atomic crowd;
They hand on their energy, loud,
Inducing close neighbors to jostle, too,
And soon the jostling disperses, too—
The irreversible change but the potion
Of the random, motiveless motions.

And such does hot metal cool, as atoms swirl,
There being so many atoms in the world
Outside it than in the block metal itself;
Entropy's statisticals average themself.

The illusions of purpose lead us to think
That there are reasons, of some motive link,
Why one change occurs and not another,
And even that there are reasons that cover
Specific changes in locations of energy,
The energy choosing to go there, intentionally,
Such as a purpose for a change in structure,
This being as such as the opening of a flower;
Yet, this should not be confused with energy

Achieving to be there, in that specific bower,
Since, at root, of all the "power",
Even that of the root of the flower,
That there is, is the degradation by dispersal,
This being mostly non reversible, and universal.

The energy is always still spreading, thencely,
Even to some temporarily located density—
An illusion of specific change
In some region rearranged,
But, actually, it's just lingering there, "discovering",
Until new opportunities arise for "exploring",
The consequences but of random opportunity,
Beneath which, purpose still vanishes entirely.

Events are the manifestations
Of overriding probability's instantiations—
Of all of the events of nature, of every sod,
From the bouncing ball to conceptions of gods,
Of even free will, evolution, and all ambition;
For, they're of our simple idea's elaborations;
Although, for the latter stated there
And such for that as warfare,
Their intrinsic simplicity
Is buried more deeply.

And yet, though sometimes concealed away,
The spring of all creation is just decay,
The consequence and "instruction"
Of the natural tendency to corruption.

Love or war become as factions
Through the agency of chemical reactions,
All actions being the chains of reactions,
Whether thinking, doing, or rapt in attention,
For all is of chemical reaction.

At its most rudimentary bottom,
Chemical reactions are rearrangement of atoms,
These being species of molecules,
That, with perhaps additions and deletions
Then go on to constitute another one, by fate,

Although, they sometimes only change shape,
But, too, can be consumed and torn apart,
Either as a whole or in part; so cruel,
A source of atoms for another molecule.

Molecules have neither motive nor purpose to act—
Neither an inclination to go on to react
Nor any urge to remain unreacted;
So, then, why do reactions occur, if unacted?

Molecules are but loosely structured
And so they can be easily ruptured,
For reactions may occur if the process energy norm
Is degraded into a more dispersed and chaotic form,
And, so, as they usually are always constantly subject
To the tendency to lose energy as the "abject"
Jostling carries it away to the surroundations,
Reactions being misadventure's transformations,
It then being that some transient arrangements
May suddenly be "frozen" into "permanences"
As the energy leaps away to other "experiences".

So, molecules are a stage in which the play goes on—
But not so fast that the forms cannot seize upon;
But, really, why do molecules have such fragility,
For, if their atoms were as tightly bound as nuclei,
Then the universe would have died, being frozen,
Long before the awakening the forms "chosen",
Or, if molecules were as totally free to react
Every single time they touched a neighbor's pact
Then all events would have taken place so rapidly
And so very crazily and haphazardly
That the rich attributes of the world we know
Would not have had the time to grow.

Ah, but is it all of the necessitated restraint,
For it ever takes time the scene to paint,
As such as in the unfolding of a leaf—
The endurations for any stepping feat,
As of the emergence of consciousness
And the paused ends of energy's restlessness:
Is of the controlled consequence of collapse

Rather than one that's wholly precipitous.

So, now all is known, of our here's and nows
Within this parentheses of the eternal bough,
As well as the why and how of it all has come,
And of our universe's end—but, that others become.

(The verse lines, being like molecules, warmed,
Continually broke apart and reformed
About the rhymes which tried to be nonintrusions,
Eventually all flexibly stabilizing to conclusion.)

THE END OF THE ROAD

My steps fell heavy on the ground,
As the dusk settled in all around,
And I passed the stones,
Where, thereunder, grass blanket overgrown,
The forefathers of the hamlet slept.

Although dead upon,
I stopped not to rest,
But plodded on,
To where autumn's last garden grew
Through the leaves that fast the fall bestrew.

With her sight, my step was lightened,
As in a hearth when a flame quickens,
For there ahead, alone,
The golden light of my angel shone
From a heart fair
That reached out in rays,
And lifted my own—

A star, lighting the way,
Guiding me—

On a night when every road
Led toward home.

REAL—>ILLUSION PROGRESSION (NOBODY'S)

Real
!
!
V

The weightless tons of massless
Of photons stream into our eyes,
Carrying to us the information of
The objects that emitted them.

Do we see a dead star shining brightly
In the fully present Now
Albeit partially past then?

Hitherto whereby whence we came?

Are we talking consciousness relative to 'C'
Or fully present as events they occur
Thence they went?

(—Nobody, and more following,
From him, too, but also from who knows where)

Illusion coming, as promised,
But note that even that
Must have a "real" basis beneath,
As must all TOES.
!
!
V

There was no time, and no space.

Even now,
Mass-energy minus Gravity equals 0.

Infinite speed is as no speed,
All already there.

"Things" are created from light,
All the effects happening in no time,
All at once,
But, brains process slower than light.

Consciousness, then, is in effect a result of
The slowing down of the speed of light
Due to centripetal forces that form the atom.

Can an absolute vacuum have the *potential*
To differentiate itself from itself?
[Yes, if it's not totally absolute.]

There would be two different
Perspectives to the state:
One would be more expansive
Than any infinite Universe we could imagine;
And the other would be more contractive
Than any point we could imagine.

Perhaps uncomfortably,
Anything and everything
Has already happened long ago
Because there is really no time to begin with—

It's all like a dream made real
[the faux true being the same as the true true]
Where individuals aren't aware of it
Because they're in it;
And, being a potential state
Based on probable outcomes,
Any dream can be made real
In relative to "time".

Yin + Yang = Tao,
Which *can* be interpreted as

1 - 1 = 0
Or
1 + (-1) = 0,

Which carries both negative and
Positive potential charge
And an infinite number
Of *different* universal states.

The Wu Chi state of the Tao or "Zero"
Would be the exact same result
As two annihilated particles.

Although,
Instead of releasing
The proportionate quanta of energy
Based on the mass of electrons and positrons,
By its *absolute* neutrality
(to the exclusion of anything and everything)
An *infinite* number of non-dimensional points
Would release an *infinite* amount of energy
Because of the division of Zero by itself —
Which is limited to 0/0; 1/1; -1/1; -1/-1; or 1/-1,
Yet can represent fractional charges
Of quarks from 0 thru 0.000~1.

Then, through imperfect replication,
Similar to DNA replication,
Information from all possible universal states
Would be recorded or encoded
And the Universe [multiverse?] itself
Could then be considered "all-knowing".

The matter is all
Attracting itself by gravity.

Two pieces of matter
That are close to each other
Have less energy than
The same two pieces a long way apart,
Because you have to expend energy
To separate them against the gravitational force
That is pulling them together.

Thus, in a sense,
The gravitational field has negative energy.

In the case of a universe
That is approximately uniform in space,
One can show that this
Negative gravitational energy
Exactly cancels the positive energy
Represented by the matter.
So the total energy of the universe is zero.

(Hawking)

If you could ride a beam of light as an observer,
All of space would shrink to a point,
And all of time would collapse to an instant.

Yet, relative matter in motion can be explained
Without the paradoxical side effects
By it being recreated over an infinite number
Of non-dimensional points,
Much like an animation on a computer screen.

There is no literal movement,
But a recreated quantity of bits of information
Gives the impression of movement.

Taken a step further,
The distance between points of "matter"
Can be explained in terms
Of time instead of space.

The Yin/Yang duality arises
From the invisible Tao,
It's positive energy creating
The virtual matter particles coming fore,
But, they are coming apart at the seams,
Because the black structure
Is a 15 year old graphic that I made
In the computer stone age,
And because negative energy
Annihilates them,
Retreating aft
As rays back to the Tao.

Come and Gone

Like the light from a star already spent,
Our 'get up and go' has long gone and went.
We all birthed, lived, and died right away—
There's nothing left but the slo-mo replay.

I'm still not sure if the subconscious mind
Is so talented to calculate
And collapse the superluminal.

Are lower species or plants doing this, too,
Or are they our phantasms?

And isn't the subconscious
Somehow hampered by being
Part of the "dream", as well, in what it forms?

The planets, plants, animals,
Raw materials, atoms, etc.,
Are phantasmal because the
Subconsciousness is universally shared.

The whole of evolutionary procession is based on time,
And it is time that is proposed as being
The sole governor of the grand illusion.

The precessional and processional
Markers are imaginary,
Like memories and predictions,
And are related to the imaginary
Dividing line of the poles I often refer to.

The mechanism is the splitting
Of the absolute state,
The unconsciousness,
Into the two abstract perspectives
Referenced in my other postings.

They serve as the basis for infinity or relativity
Because there are an infinite number
Of fractions between them.

You're right about the subconscious
Not being talented enough.

It is one-sided with regards
To "this side" of the universe,
The other being the parallel antiverse,
The correlation being the positive
And negative infinity/subconsciousness
On both sides of the absolute/unconsciousness.

So to be talented enough,
There has to be a meeting of the minds,
So to speak, and the one side has to
Pass through the absolute zero point
Which I refer to as the gateway to the way... TAO,
Or whatever name you prefer
According to your particular faith or fancy.

You've proven to me from enough indirect analyzes
That you get the correlations.

I have tried to make things real
By considering various mechanics
That could be applied,
But none are as yet applicable
In order for there to be a literal break
In the symmetrical zero state.

Though there is no difference in feeling
Whether or not one feels due to literal stimuli,
Or due to an illusory change in quantum states
That are interpreted as real.

Aside from that, Lloyd's contraction,
RascalPuff's expansion,
And Geoff Haselhurst's finite wave front
Can apply as logical mechanics.

For those who require mathematical
Interpretations of the mechanics,
All that is required for the absolute
Unconscious potential state to be split is "nothing",

And this where Fredrick's binary math
Can be applied to subconscious infinity.

The only factor missing,
Imo, is the absolute 0/0
Which is naturally the
Only choice non existence has—
It is a nonexistent abstract divided
By nonexistent abstracts to render
Our geometrical x-y axes of which
The potential state can realize
The probability states of infinity
Along those imaginary axes.

Said more than enough I think,
But just one more thing about your image...
You can't have your cake and Edith too.

The Procession of the Constituents of Reality

Sad Yesteryear, Forever, and Everywhere,
They all came—to weep for Nobody Nowhere,
With Why and How, Then, Now, When, and What & Where,
Even Inward and Outward, with Potential—
Led but by their tears and sorrow. Your posts zing
With things that "none" can bring: Everything.

The Interlude of Life, Interspersing the Real With Nobody's Illusionary

The zephyr faints, dying in the half-light,
Its caress suspended, as day kisses night,
When, for some instants, stretching into moments,
We are neither here nor there, but in twilight.

We live at this boundary of day and night,
Our selves merging in the blend of twilight:
You and me, me and you; yours, mine, and ours—
The day-gold melts into the jeweled night.

Our parentheses in eternity
Flashes as a twinkling, but's extended
By time into a phantasmic life dream
That's existent the same as if it were.

As living pearls we're strung out right and left,
Lovely and beautiful on the Earth's breast.
Her bosom heaves, as one by one we're cleft:
A thousand truths die, until none are left.

A life dream's like a rainbow, not really there,
A false phenomenon become tangible
Through its being, the true true of the faux true,
Molding a genuine significance.

Say "Farewell!" Heaven's promise is bereft;
Yet, live with gratitude—be not distressed;
Still, dismiss immortality's dream;
Accept, with appetite, whatever's left.

Life's indeterminate or not, the same
Being brought by the virtual as the true,
The mechanics being as incidental
As why "color" chose its wave frequencies.

I've said "Good-bye" to the dream of forever,
'Though I'm too philosophical to be bitter.
Poignantly resigned, I accept, with hunger
And joy, all that's left—whatever—with pleasure.

Life's here, like a virtual particle
Born this side of an event horizon
Of a Black Hole, realized by its presence
In the realm of what's been radiated.

All's right with a world without the angels—
Human, we try, we push, we climb, we lust,
We dance, we dream, we feel, and love with zest—
Yes, all this, thanks to the beast within us!

There is no difference of what makes none;
Realism is now playing, the living film:

A reality show in the theater
Of the mind's eye, with the 'I' observing.

So, I drink-in the pleasures of creation,
For what else could be the point of cognition,
If not to absorb all that comes streaming in?
Life's sensation is the main attraction!

At first it was like a moving picture show,
Attended by mysteries, row upon row,
That were faceless, laughing, in the dark below;
So I laughed, too, and better enjoyed it so.

Nature enters along paths sensory,
As it seeps into rationality,
Then saturates the being with delight—
The greatest taste is of reality.

Opposites are just a different view
Of one fundamental phenomenon—
Light, beauty, and goodness are the inverse
Sides of darkness, ugliness, and evil.

Strive for a dynamic balance—of light
And dark, Yin and Yang, and wrong and right.
Reality is found not in separate actions,
But in related events blended in twilight.

SPRING SONG

When the spring tides fill up the free spaces
In our winter spirits, then we shall roam
At ease, drink the sweets in every flower,
And feel the balm in every breeze; for we
Shall all thread the lovely web of life that
Affection's hand weaves so fine about us,
Drinking up deep draughts of life's delight.

THE 4TH YOGIC BODY...

You just have to love this guy!

Yogi Berra,
The great New York Yankees catcher,
Said many sayings that
Seemed to make sense,
But really didn't, or maybe they did
Like about a restaurant
"No one goes there anymore;
It's too crowded.",
And, about a ball field
"It gets dark early out here",
Plus "If people don't want
To come out to the ball park,
Nobody's going to stop them",
"It's deja vu all over again",
"I didn't really say all
Of the things I said",
And many more great unsayings
Just as meaningful or not
As some that we see on TQ.

In his 3rd Yogic reincarnation,
He was a coach and a manager,
And is now an elder statesman in his 4th.

Yogi Berra was simply nicknamed "Yogi"
Because he looked like one;
Nor did he disappoint in this area,
But came through time and time again
With his enigmatic observations.
This was his overall Yogic self.

< 7 >
— 3D Glasses —

Hard reality is constructed by our brains,
From some kind of frequency domain,
As when sound waves turn into sound, so, the
Apparent physical world is an illusion.

POSITIVES

Success blossoms out of a thoughtful dream,
Grown from seeds of what life to you should seem—
Then bears forth fruit healthy and delicious,
In the garden watered by a wishing stream.

In the night lies the healthy breath of morn;
The giant oak sleeps within the acorn;
The flower waits for spring inside the seed;
And so in a vision is one's life born.

Think positive and thus conquer the strife:
Think of love, health, beauty, adventure, wife.
Beliefs manifest themselves in reality:
If beliefs are halfhearted, then so is life.

The revelation hit me like an hourglass,
One made of the heaviest welded brass—
I rose with a start, my life now begun,
Before more time through me unaware could pass.

In the darkness I alit from the Wiz,
And tried to make sense of this world of His.
Now I've found the answer to life's dark quiz:
One must live this life by what light there is.

I've constructed the world that dreams require,
One moulded closer to the heart's desire.
In this world-body of a soul inspired,
I'll live life entire before I expire.

My blood runs warm with the sun's heat at noon.
My spirit is swept by the swelling moon.
Air surrounds me. The ocean flows through me.
Earth's rhythm is always playing my tune.

I pursue the shadows of forms that live
In dreams—perfected ideals that outlive
All the minutes and hours that time devours.
I seek what hope creates, what wishes give.

'Twas a time before birth when we were not;
T'will be a time again when we are not.
From Death our life was a borrowed debit—
We spend it, love it, and live it to our credit.

Visions pour forth in positive images,
Thoughts creating life of former mirages,
Ideas developing from the negatives—
Life's emergent dream ever encourages.

Never struggle against the way things are,
But rather, become the way that things are.
When you give yourself to the moving whole,
Natural currents will carry you quite far.

No ego/self—just interrelation—
Life's oneness is a complete sensation.
It's beyond intellectual concepts, thus,
It defies any further description.

GREAT MINDS

Our great minds have come from the mindless,
For then there is no regress of
...MIND having to form Mind to form mind,
Have deduced the nonspecific, indefinite,
Eternal, causeless ground-state
That produced its arbitrary and local specifics
With definite forms that worked,
No time being special, since any time,
No place as specific, since anywhere,
No special laws, since many possible,
No form right or wrong but only
That some forms may be
Stable enough to continue on,
Into complexity, since they
Were also flexible enough not to be inert.

OUT IN THE REAL WORLD

We were in a bookstore the other day,
Summaria and I, in the Science section,
It being on the left,
With the Dogs and Cats section on the right.

A redheaded lady, finely dressed
Was sitting on the floor, reading 'Antimatter',
Trying to find out what particles are,
For we inquired of her.

A kind of 'mad-scientist' then arrived,
Looking for a Science Dictionary,
His hair much worse than Einstein's,
Plus, he was all shabbily dressed
With really baggy clothes
And had probably gone without a bath for weeks.

I asked him if he was a scientist.

He said "No; I would be, but the pay is not good."

Another lady appeared,
Looking for the 'Poodles for Dummies' book.
Someday, all bookstores and libraries
Will have to double in size to hold
All the 'for Dummies' and 'for Idiot's' books.

As you can imagine,
Some karma spread unto these people
And we were soon all sitting on the floor
And having some kind of informal class
On Anything and Everything.
I had become sort of
A Professor for Science Dummies.

"What's a particle?"

"Electrons, neutrinos...
You can find them on the internet
Under 'Standard Model',

So, don't waste money on a book."

The redhead wrote this down,
Along with everything else that got said.

"Maybe God is the particle," she offered,
"But where did the antimatter go?"

"Maybe a lot of it glommed together
And went down a Black Hole or something.
At least we have mostly uncle-matter around here,
Thank God."

"Could God be the particles?"

"That sounds very restrictive,
As well as being a lot of information to manage.
Let's just let the particles be the particles,
And simply have them do their thing."

"Oh. Well, energy comes from the stars and planets.
I've kept a log and there are different effects
Depending on the time of day,
Plus those energies, of when I was born."

"Astrology?"

"Yes."

"Well, the doctors and nurses
Surrounding your birth
Would have had far greater effect
And influence on you at birth
Than some stars and planets far away,
Although they do emit some amount of energy."

"They determine our lives with that energy."

"Yes, true, there is energy
But I don't think stars and planets just sit around,
Thinking 'What should we do to this guy;
What should we do to that person'."

"They decided that I would get hit
By a tractor-trailer truck, which nearly killed me."

"Wow! Glad you made it."

"When it happened, I had no memory for a while,
Being somewhere else in another dimension
In some blank space."

"Let's just say that you
Got the hell knocked out of you."

"Could be."

The shaggy hair guy was listening, too,
But was getting perturbed
At finding no 'Dictionary of Science',
But the Poodle lady was taking it all in,
Never saying a word.

"There has to be a cause for life, professor."

"Causes of LIFE making Life making life
Can't go on forever; so, no go on that one."

"Ever see someone turn into light
Right before your eyes?"

"No, you?"

"Yes, and these are like spirits and angels."

"Then it returned to normal?"

"Yes. And the spirits are here right now."

"I don't see any; hey, who bumped me!"

"You can't see them, but they're here."

"Hey, where's Summaria?"

(She had played a fine joke on me, skipping out,
Leaving me stuck with all this hocus-pocus stuff.
Even the 'mad scientist' had gone,
Looking for a store clerk.
All I had other was the Poodle Lady.)

...A bunch of really fine talk flowed
That had a lot of good stuff and jokes
That I have put on TQ, at times,
But, well, you had to be there...

"Are you a Buddhist? You sound like one."

"No, for they believe that all is illusion;
Otherwise, fine, as they serve the task,
Like always picking up litter,
Not worrying if anyone is watching;
Although they don't have a God,
But just a human guy, Buddha."

"All could be a dream, such as us being here now."

"I've heard this one.
I knew the Great Lama
Of the Eastern United States.
He owned a restaurant near the train station,
And I got to know him pretty well,
His bodyguards retreating.
He even offered to take me to India with him,
But I stayed here."

"Any great wisdom?"

"Yes, for when I asked him
Who really won the election,
Bush or Gore, he said "Who cares!"

"A very great wisdom, indeed."

The bushy-haired guy reappeared,
With a store clerk;

They couldn't find a 'Science Dictionary',
And so the guy got mad and left.

The store announced that it was closing.
Summaria peeked around the end of the aisle.

The redhead offered ,
"I'm inviting you guys
To have dinner with me and my friends."

I looked over at Summaria.

Summaria said "Great; we'll go."

...

And a fine dinner, it was,
With much further and good discussion,
That which spurs even more thoughts,
Such as on ToeQuest, in a fine mansion, no less,
The redhead telling us we could stay
As long as we liked and/or visit her,
Whenever, coming and going.

And such the karma was flowing,
So we are having a wonderful vacation from our tent,
There being servants and all.

THE UNIVERSAL JAZZ

Jazz, seeming like a bunch of noise at first,
Which it is, then finds its tune,
Like the TOE that entangles and entangoes
Into the one of the dance and the dancer.

THE REAL WORLD CONTINUES...

The redhead looks like a rail, frail and thin,
But energetic nonetheless.
She is really old [looking], being 60,
Which, I know, is younger that I am.

Too much sun, perhaps,
Plus there was that tracter-trailer accident long ago.
She is a combination of
The dreamy but fun Melanie
And a positive gleeful Mikal
But having an open
And wishful scientific-to-be path,
One that comes and goes vs.
The non-conceptual invisibles.

Rare that a caucasian owns a plantation,
But her deceased husband was oriental.
These plantations, of which there are many
In the level interior of the island,
Raise cane and whatever else in this fair climate.

Kind of a laid-back atmosphere here,
The workers coming and going with ease,
Even into the glorious white mansion
And its outbuildings.

We can see the back and the side
Of our mountain way off in the distance
As it calls to us from this lowly point,
And so we come and go.

The cook here is fantastic,
Blending all sorts of seafood
And vegetables with the hardier stuff as well.

Protons and wantons abound here as well,
Making me think of ProfPat.

SOME OF MY RESEARCH ON THE "DEVIL"...

— Myth —

The eyes love to rest on the sky of blue
While Eve upon the greensward smiles at you—
A new life colors the world in between
Devils and Angels: Earth's human pristine.

— The Bad Old Days —

Life's still emotionally primitive—
Negative feedback mechanisms in
The central nervous system, now useless,
Still send thousands-of-years-old messages.

— The Fall of Evil —

Emotions are but molecular events,
Some forced upon us all, like jealousy,
And some others, like aggression, born from
Low serotonin, NOT from the Devil.

— The Root of Ill and Evil —

Low serotonin stems from genetics,
Stress, lack of exercise, or the wrong foods,
And can cause anger, anxiety, and
Depression, even bad behaviors, crimes.

— Chemical Imbalances —

Since the aggressive urges leading to 'sins'
Are not caused by (D)evil, the 'sinners' are not
To blame, although we still have to lock up
The violent ones to protect ourselves.

— The Devil is Dying —

Six hundred ago, the church thought that ills
Of a physical nature were caused by
Evil spirits; however, now we know
They're from bacteria and viruses.

— The Devil is Dead —

Now the church thinks that ills, or sins, of a
Mental nature are caused by the Devil,
An evil spirit; however, now
We know of brain chemistry gone astray.

(And poor upbringing,
And the sometimes spurious molecular events
Of very extreme and bad emotions, etc.)

SENSE OF SPRING

In winter's sleep I lay enrapt,
Safe in a chrysalis, thought-bound.
Oh, pleasures of the mind! Enjoy—
They're intensely sweet to the depths.

Dreams rival the sense's pleasures—
They're alive even in the mind:
Within is as real as without;
Within is where senses make sense!

When I wished and dreamt fantasies,
Life lived in my sanctuaries—
Influenced not by the senses,
But by soul, heart, and memory.

Such waking-dreams must ever be fed
If they're going to grow without. So,
As spring returns, dreams take wing—
The senses reign; the mind can rest.

UTOPIA, WITH NO MYOPIA

We slept over ten hours last night, a fine luxury.

Another easy going day day
On an utopian island
That is racially democratic
(We are a minority here),
The temperature hardly ever going above 81,
Or not even above 84 in the "summer".

The motorcycle idles,
Which is really nothing more than a soft purr,
As one of my long-life laptop batteries charges up,
Waiting to be swapped with the other,
Soon to be dying one.

There is swimming in the ocean
In the late afternoon,
Whenever, then tennis, some days,
Followed by a dip in a
Freshwater lagoonish type of pool,
Then Smorgy's buffet,
Now and then, as a treat,
Savored outside on a
Lanai overhanging the ocean
And under the stars,
For night falls at 6 PM or so,
The sun even plummeting
Like a deadweight,
Relatively speaking,
With but a short twilight thereafter,
And, yet, the sunsets are often glorious ones,
The colors more rapidly changing.

Then closeness, later, in the tent,
And wonderful sleep...oh, beautiful sleep.

My partner will be posting on TQ,
Although perhaps not so often, but who knows.
It will appear under my ID
Since we only have one computer,

But I'll prefix hers with "Summaria:",
As she is good at summarizing ideas,
Whether good ones or not,
Plus, of modes of being
And so her posts may be short and sweet.

Summaria:

Aloha, TOE!—as it, too, both comes and goes [on].

Eat, sleep, play, drink, nature, love, thought,
And sex are what we are made of,
A rather beautiful meld beheld.

We know that self organization must be so,
Else we could not have been here to know.

Throw the rope to the drowning ToeQuestors,
So that life and ToeQuest may go on,
Whether by their hanging around
Or hanging themselves;
Here, where anything goes.

If the supposed "God" cannot be proven
Then "God" is no matter [for what matters.]

Dearest me[l]—
A self that negates itself is an oxymoron.

If Brahman is as real as real can be
Then so are we [in the lovely zombie jamboree].

Here's a good, but solvable mystery:
Look at your eyes in a mirror,
Trying to see them looking and moving,
Even far to the left of right…
You don't see them move at all,
But only being still,
Yet, we can see other's eyes move.
How come?

There is a problem with this
Highest Brahmanic order
Occurring right off the bat,
And an infinite one, at that.
But to each (if there are), their own
Live and let live.

Ah, that old Devil, Brahman,
Even puts illusions of illusions
In the optical illusion opticals of 'us'
That aren't there either.
I'm beginning to lose my trust in this character.
The eyes have it—the ayes,
And the Scientific American articles
Demonstrate the real brain
With its real visual systems.

Good feelings arise even further
With moderate warmth, more or less,
Plus an outlook of joy;
Cold ones more so of the cold
Or of the very serious, although not always.
People do not look down at the ground here
When they pass you.

Negatives, humorlessness,
And an overly serious and very fixed outlook
And attitude can ruin the fine recipe
Of luck making its own [good] luck
Of 'karmatic' successes.

'+/-' can replace '='.

< 8 >
— Movie Starring —

The virtual reality can be
Enjoyed and directed in lucid dreams,
Where one can do anything at all, without
Injury or penalty, with real feel!

QUALITY TIME

When push turns to shove,
I turn to love,
And retire to our woodsy home,
Far from the noise of day,
Where my partner and I
Live and laugh and play.

Here the phone cannot ring,
And there are no bills to pay;
We drench in the joys of morn,
Pulling out each other's thorn.

There are no deadlines to meet,
No interruptions allowed;
We wander each other's way,
The hours to drink away.

No hustle-bustle or rushing
Hither and thither
For this and that
Small and smothering detail.

We just kiss away the stress,
And together caress,
And cuddle in the
Flickering candlelight—

For, in the love temple,
There is only Now.

< 9 >
— Speedy Searches —

A simple four-way lookup senses taste
By degrees of bitter, salt, sweet, and sour,
And, likewise done, the three-way colors,
Plus ten-plus-way facial recognition.

THE TRIP

...It's wonderful to see that somebody
Has recognized the importance
Of Emotional Intelligence and that spiritually
It is a higher field of energy.

I know from my own life experience
That emotion refined and as a field
Is the ethereal boundary between
The physical and ethereal realms.

As far as I know from my studies
This is the boundary that
Higher creational beings
Not taking physical manifestation
Can manipulate and interact with
Our physical realm and us as souls.

I knew in the experience that I was not crazy,
Was not imagining the experience
And this was bolstered by the fact
That I have always been a nondrinker
And had no experimentation with drugs
Which could alter my consciousness....

The key here is still an altered brain state,
If this was during a likely NDE or an OBE,
Which are of extremely altered brain states
That are certainly not the normal operation.

Being drunk or on drugs would be
A relatively "sober" state compared to these.

I had an OBE in which
I even saw dream images still moving
While I was becoming half-awake,
These being overlapped with
The real reality of me and my room.

Thus, it is not at all likely that
The ethereal higher creational beings

Exist that were sensed as real,
But simply visions seen or felt
During an upset and altered brain state.

WORTHLESS AND PRICELESS

The poet works only for love,
And for nothing more.

There's no profit in coin,
No wealth below or up above,
No fortune told,
No living made.

The poet writes only for love,
For there is nothing else:
Just a few readers, and
No business worth speaking of.

Yes, I know that I'm no bard,
And that fame is only met,
If at all, in the graveyard,
Where, far beneath,
I cannot grasp the laurel wreath.

As a poet, I write much of love,
Of it's worth and wealth
Measured in goodness
And beauty seldom heard of.

Without promise, the poet writes on,
And knowing well
That there shall be no award,
But, ever on the poet writes,
And lives, and works for love,
For he's found that love
Is its own reward.

BECOMING

We humans mirror and recapitulate
All of evolution while growing in our mother's womb,
Racing through the stages in which life evolved.

During this nine months, and even beyond that,
We move from mindlessness to shadowy awareness
To consciousness of the world around us
Onto consciousness of the self
And then even to becoming conscious
Of consciousness itself.

For the first two and one-half years of life,
The inexplicable holistic world
Is experienced less and less holistically
As the child discovers the bounds of discrete objects.

The holistic right brain remains, of course,
For us to take in the overall view,
While the logical left brain is also there to recognize
The detailed relationships between objects.

As such, so goes the universe,
Since we are formed in its image.

So, then, this gives us a clue
To the nature of the universe.

Seeing that the brain is
Divided into two hemispheres,
Each with their own
Characteristic mode of thought,
That that can communicate with each other—
Means that we are looking very deeply
Into the way that reality itself is constructed.

These two complimentary aspects
To the cosmos are thus absolutely essential,
One being of the whole:
The apparently indivisible, continuous fluid entity
[Although discrete at unnoticeable levels],

The other being the interrelationships of the parts.

Each interpretation cannot appear
At exactly the same time,
But the Yin ever gives way to Yang
And ever then back to Yin, and so on,
The rounded life of the mind
Thus continuing to fully roll
As the cycle of this symmetry
Turns and returns;
If not, one either gets totally lost
In the details or prematurely halts
After but an apparent whole.

The holistic right brain mode is unfocused,
As we see in some people
Who are unconcerned with details,
It always building the scene in parallel
To form a single entity;
Whereas, the focused left side of brain
Isolates a target of interest and tracks it
And its derivatives sequentially and serially.

Yet, the two sides of the overall brain
Are connected to each other
And so the speed of the juggling act
Can meld them together
Into a complete balance like that
Portrayed by the revolving Yin-Yang symbol,
Each ever receding and giving rise to the other

Such does the universe go both ways, too,
Its separate parts implicated
With everything else in the whole.

During conscious observation,
The 'hereness' and 'nowness'
Of reality crystalizes and remains,
We establishing what that reality is to some extent.

We define and refine the nature of reality
That leads to the mind's outlook.

Counterintuitive? Cyclical?
Yes, but it is the universe in dialog with itself;
The wave functions, and yet the function waves.

The universe supplies the means of its own creation,
Its possibilities supplying the avenues
And the probability and workability
That carve out the paths leading to success.

So, here we are, then and now,
The rains of change falling everywhere,
The streams being carved out,
The water rising back up to the sky,
The rain then falling everywhere,
The streams recarving and meandering
Toward more meaning and so on.

RANGES

On the scale of atomic nuclei,
The weak and the strong forces dominate,
But are limited in range.

Over larger distances, e/m takes over,
Its interactions accounting for almost all
Of the chemistry and physics of materials;
However, charges come in both
Positive and negative forms,
And so they tend to cancel
Into neutrality for large objects.

So, on the very largest scale, gravity dominates,
Although it is the weakest of forces,
For it stretches over a long range
And always accumulates,
There being no such thing
As a gravitational charge.

Nor does dark energy count as “antigravity”,
Since its energy density is positive,
Even if it exerts negative pressure.

WILD ROSES

I cultivate wild roses in the spring.
Do I try to tame them, breed them, subdue them?
No, I encourage roses wild and free—
The wildest plant's the one that's most alive.

Her leaves are coarse and brown at the edge,
For they contend for life near tree and hedge.
I develop wild roses in the spring—
Such we taste the first sweet breath of summer.

I hold thorns in my hand; I hurt, I bleed.
But I don't let go; I hold her tighter.
The winds come, the rains fall, the storm passes;
The breezes caress her—she blossoms forth.

Now she unfolds to all, and so discloses
That there is new life among the roses.

MADNESS TO GLADNESS

Behold the Madman! He finally cometh,
After all of these years, complete at last,
As he savors the gladness of life—
Upon all worlds his shadow is cast.

Emotion rules, tempered by intellect,
For he fights sadness with madness;
No wait for the world—he comes to it.
Behold the madman in his gladness!

He bites the dust, only to rise again
Through endless skies with every yen.
He directs the madness of his essence
With a shade of thought now and then.

While accepted thought and doctrine
Are madness, and madness is truth,
He loves as no man loves emotion—
For he's met the world and now it's his!

COMPLETING THE OTHER

Men and women can't exist in isolation,
For, like valleys that give rise to mountains,
The nature of one makes necessary the other—
When they're joined in love, there's wholeness again.

The Warning:

So, then, what about the state of matrimony,
Hmmm... I mean monotony, no, I mean monogamy,
No, I mean the state of confusion,
Because... opposites attract, at first,
But then the friction grinds on one's nerves
So, perhaps one then looks more for similarity
And compatibility in the years thereafter.

Now, you can, in theory,
Have two or more spouses
(The plural is not really spouses
But is actually 'spice'),
For, while there is not much wrong
With giving love to many,
There is not usually the time for that,
Except for many elderly Mormons of certain sects
Who have several teen-ages wives for certain sex.

Presumably, this marriage won't last...
So, then, many years later we hope
That you will still be nice
Enough during the divorce
To give your ex the gold mine
Rather than just the shaft.

Going Ahead With it Anyway:
(Via a comic book Minister)

The marriage vowels go something like this, I believe,
After some translation; please repeat after me:

"I, Miss-Understanding, take thee,
Mister Right, first name Always,
As my awful wedded husband,
Through thickness and thin,
For butter and for bratwurst,
But only while in richness and in wealth—
Until our debts do us part,
Bedding thee with these marriage vowels:
Aaa Eeee IOU! Y(WHY?)."

"I do. (which is the longest sentence)"

(hot kiss)

Your husband has now
Given up his BS Bachelor's Degree
And has hereby granted you
Your M.R.S. Degree,
The engagement ring having led
To the marriage ring
And thence to the suffeRing—
That pre-divorce state of union
That often at least goes
From here to maternity
To his absence making
Your heart go wander.

Attraction

There's an urge between root and flower,
Plant and soil, leaf and sun, air and water,
Daystar and planet, valley and mountain,
Wind and mist, man and woman—for ever—

Bubbling Passion

Like water, a woman is slow to boil,
But, likewise, slow to cool down afterward;
Man, like fire, can be ignited and quenched;
Yet, fire and water in balance make steam!

THE GOD THEORY

Alfonso Rueda, a German physicist
At California State University
Grinded through a complex mathematical analysis
Of F=ma which to him worked out
A radical new understanding
That inertia is a fundamental property of matter.

Rueda derived that mass
Becomes in effect an illusion.
Matter resists acceleration not because
It possesses some innate property called mass
But because the zero-point field exerts a force
Whenever acceleration takes place.

Put in metaphysical terms,
There exists a background sea of quantum light
(The electromagnetic zero-point field)
Filling the universe,
And that light generates a force
That opposes acceleration when
You push on any material object.

The action of that quantum light
Is what makes matter the seemingly,
Solid, stable stuff of which
We and the world are made.
(The God Theory—Bernard Haisch)

This is kind of like the supposed Higgs boson,
Which, while also called the 'God particle',
Has only to do with physics, not "God".

Some wishful people are ever and always
Trying to find some space for God
To have done or be doing something.

The latest and greatest news of the LHC—
The Large Hardon Collider—
Is that, just a day go,
On April 1, 2010,

Its energy output really spiked,
Even tripling its supposed maximum.

Unfortunately, the LHC blew up.
Exploding with the astronomical energy
Of a aircraft-carrier-size monster truck,
Destroying all of Switzerland.
Even all of its alps, along with
Most of the secret money of the world.

From an e-mail that escaped the LHC
At the very last nanosecond
We have learned that, indeed,
The Higgs boson was found,
And, so, as the LHC
Has really already done its job
It wouldn't have been needed
Any more anyway.

Seeing you have more than likely
Not read this scientist's book,
You would be standing on a pretty shaky pedestal
To judge his summations.
Also because your statements
Rise from bias and assumptions,
It would be shaky indeed!

Nope, for I have disproved God
Even in several ways of self-contradiction,
As I've posted on TQ.
Thanks, though, for putting me on a pedestal.
Only the believers make up things
Out of an invested interest (bias).

From the particular flavor of the book,
I think Haisch would tell you...
"Live and let live" and "scientists should stop
Arguing and wasting their energy,
About something which can
Be neither proved or unproved."

He did state in his book that
The Scientific reductionist view does not uphold
A purpose for life and as humans
That might be bad news and also information
That undermines the ethical
And moral underpinnings of society and civilization...

God not existing is merely
A byproduct of doing science.
There was no purpose for life,
The ground-state being causeless, thank not God,
For this truly allows the “free will”
Of developing one’s own fine and good purposes.

Moral underpinnings are not undermined at all,
As they came from humanity itself.
Mythical pinnings are not useful for reality
Since they are not of reality.

He is a scientist, makes that clear,
Defines what he cannot adhere
To such as intelligent design
And then creates a God theory
Relevant to what may possibly be behind
The Big Bang and the fields that define the universe....

Yes, ID and other notions substituted
Keep falling like dominoes,
There being retreat after retreat.
The last domino shows snake eyes.

Behind the Big Bang is just more stuff
Until we get to the causeless
That forever had to be
Since nothing can become of Nothing.

Look for Great Mind
In the further complexity of the future,
Not in the simpler and simpler past.

THE CONTRADICTORY CREATOR
(The Disproof, in Short)

A Creator's
Mind-Life cannot be
First and Fundamental,
For parts precede a system.
Done. Go on, just for the heck of it.

Nothing begets nothing;
Thus, something was always here;
So, no creation means no Creator.
More than done.
Go on, just for the heck of it.

Only the natural is seen,
Nothing super.
Really really done.
Go on, just for the heck of it.

If one still wants a God...
Must prove.
None.
Go on, just for the heck of it.

Make up stuff.
No go. Go on, just for the heck of it.

God is sensed.
No go,
Due to all of the above,
Plus, many substrates beneath

Still preach God as fact.
Unethical.
Only a conjecture.

Get mad/angry.
Useless—shows nothing.

Still stubborn?
Do some science.

WHY THERE'S SOMETHING ALWAYS
(Plus the Super TOE)

Nonexistence
Of something existing
Is absolutely impossible,
And it always will be,
As it was ever was,
Being perfectly unstable,
This motion never ceasing.
It being the prime mover.

Of this instability of the simplest
The QM fluctuations must become,
All their emissions summing to zero,
Yet, a few of them somewhat enduring,
In their canceling oppositeness,
That must then on to make the definites,
So as to gain more and more stability,
Structures which must then be tied to time,
These growing more and more complex,
Even unto stars, galaxies, universes,
And life forms, such as homo sapiens,
These structures of the total cycle
Yet still decaying into their return.

Such has it always been.

From this notion (TOE) we can also dispense
With the paradox of a specific amount
Of definites having always been around
With nothing prior to account for the amount,
As well as no "where" being special either,
Since anywhere—being any old place,
Nor "time", since for any time
And for all time,
Nor "what",
Since anything
That's possible can be,
Nor "who", as special,
For any life may form,

Nor any particle properties special,
For only what is stable enough
To continue, yet flexible enough to change
Will go on and on into recombinations
Of further composite complexity,
Nor, of course, any planet,
Solar system, galaxy,
Or universe being special,
As there are any and many.

USAGES OF ATOMS

In stars, the simpler atoms melt,
They then reforming into higher ones.
Spewing out as all the atomic kinds

The atoms forming life come and go,
Which is of little consequence
Since atoms of a type are all the same.

Atoms entering to be within life
Play a part in metabolism,
The very hallmark of what is alive.

Harder structured materials or machines
Also endure, using another method,
That of mostly keeping
Their original atoms intact,
Any stray atoms adjoining
Having but little effect.

WILL

Wider learnings and experience
Allow for a wider range of actions,
Yet always and still reflecting
That which we have become.

KNOWLEDGE ABOUT REALITY AND OF REALISM: Knowing Where There is and Where There is Isn't And How We Know

That we only see the inside of our brain
Is certainly true and is also fun to explain,
Our senses detecting some of the emanations
And forming representations of them—
Even painting a finer face upon them.

As for our knowledge of the reality that's out there
We have a very excellent understanding there,
For, not only do our instruments measure it bare,
But we use the data derived from it out there
To make great models and formulas,
Which in turn we utilize to design and construct
Many intricate and working devices out there
All made from material that's out there—
That precisely depend on what's out there,
And, so, this proves that we know reality,
Even its quantum mechanical nature, really.

Is there really macro-realism, then?

Yes, it is for certain that there is macro-realism,
Meaning that it has pre-existing properties;
For example, your couch's properties,
Such as its color, remain the same
No matter how and when you observe it, sane;
Thus, the moon and the earth are still there
Even without one's observing stare;
They are even seen to have progressed,
As some things have changed or moved.

Now, what about micro-realism,
Such as for quantum entities?
Is it true that there isn't any?

Well, it's no surprise that there's no realism there,
That is, there are no pre-existing properties,
Such as polarity, spin, and position, and more,
As was always thought to be the case;

For, recently, it's been proved all the more
(Bell's previous tests only having shown
That either locality or realism was wrong
But not knowing which one it was),
The first IQOQI test being 3 orders of magnitude
Against any realism being within the quantum realm,
With a second, and, much better, test resulting
In an amazing 80 orders of magnitude against.

So, it is still that quantum entities
Have no pre-existing properties,
Just being a superposition of possible properties,
Which, too, was expected, as well, due
To their necessarily causeless realm
With nothing prior possibly there
To dictate any specific, definite state,
They being completely outside of all cause,
As they must be—where the bucks stops;
For, what good would but more possibility be,
Plus, regresses could not reach infinity.

This is again bad news for the Super-Potions,
Those ever wishful and imagined Notions.

Large, macro entities "wave-collapse" in an instant;
But, micro, quantum entities remain in limbo's state
Indefinitely, unless they interact or are observed,
In which they randomly gain definite properties.

The other entities, which are those in between,
At the boundary of the macro and the micro,
Such as, perhaps, a spec of dust, settle down,
Spending less than a second dancing around.

So, only beyond the causeless quantum realm
Does then the "cause and effect" come to be
And, then, whatever happens has direction,
But for any more quantum happenings to them.

Now, why do the quantum's
Vibrating fluctuations have to be?

It is that nonexistence cannot be;
Thus, something has to be.

There having to be the simplest
Something is natural, never the less;
And, also normal, but unguessed,
Surprisingly, is ever motion, not rest.

THE RANDOM DANCE

The simplest something is natural, never the less.
Being normal, as well, surprisingly, is motion, not rest.

The quantum jitterbugs arose, in a dance,
As they are ever wont to do, perchance,
Because nonexistence cannot be a reality;

These fluctuations create particles freely,
And their antiparticles, that eventually
Cancel back to zilch, or near entirely;
So, in a way, all that is "reality"
That is somewhat duringly
Present now is but an expression existing
Of Nonexistence's absolute instability,
Or at least of Nonexistence's impossibility.

One could say, then, that all is for naught,
(And of), as it is involved in the sense bethought
That "nought" is so perfectly unstably wrought.

And so are those fluctuation's simpletons
Emitted not so stable, becoming undone;
For simple things seldom remain unrearranged,
As they go through phase changes
And recombinations, and so forth, as deposits,
Onto and unto the more complex composites.

ZERO

In quantum theory,
Particles can be created
Out of energy in the form
Of particle/antiparticle parts;
But that just raises the question
Of where the energy came from.
The answer is that the total energy
Of the universe is exactly zero.

The matter in the universe is made out of positive energy.
However, the matter is all attracting itself by gravity.

Two pieces of matter that are
Close to each other have less energy
Than the same two pieces a long way apart,
Because you have to expend energy
To separate them against the gravitational force
That is pulling them together.

Thus in a sense,
The gravitational field has negative energy.
In the case of a universe that is
Approximately uniform in space;
One can show that this
Negative gravitational energy
Exactly cancels the positive energy
Represented by the matter.
So the total energy of the universe is zero.

Now twice zero is also zero.
Thus the universe can double
The amount of positive matter energy
And also double the negative gravitational energy
Without violation of the conservation of energy.

"It is said that there's no such thing as a free lunch.
But the universe is the ultimate free lunch."

— Stephen Hawkings

THIS TOTTERING EXISTENCE

So called "empty" space is vital,
For that's where there's the recital
That forms and plays the tunes of reality,
The grand cosmic symphony—
As existence fluctuates with the non,
Those causeless waverings of undulation.

It was once thought that the shove
Of this total energy was of
The order of 10**120 orders of
Magnitude above.

Well, if that were so near,
Then we couldn't even be here;

It was the worst calculation
In all of scientification;
So, we weighed the universe,
Summing all of its constituent verses.

The universe weighs nothing at all!

This, too, since we found that
Our universal space was B flat—
Not just via the 60 degree angles
Of a very small triangle.
Not even using stars,
One that went from here to Mars
To Venus and back,
But all the way back
To a degree of the CMBR,
Which represented 100,000 light years,
And measured the curvature:
The rays didn't converge or diverge.

The ultimate of this geometry
Is that being flat is a beautiful symmetry
That leads to yet another beauty: zero.
The ever returning, conquering hero.

Far from being the *Magnificat,*
We are more insignificant
Than we ever imagined, even Kant,
As all is a big nothing,
But also, since, considering
That all the specs of matter's amount,
For whatever is the measly count,
Compared to dark matter and dark energy
Are but a kind of pollution, irrelevant, really.

DEFINITION

So, if a definition is indefinite,
What defined it to be so?

Can't have a definition with
Nothing defining the causeless.
The indefinite requires no definition.

Yes, I'd have to agree that sequential logic
Doesn't appear capable of computing things
From the no-beginning of forever.

The sequence from back then
Could never have completed
Its forever unto now.

It defeats purpose, of course,
But this provides freedom.

The How is the chaos of simplicity
Recombining towards complexity.

We are all of the stardust that came
From hot young stars, and more,
That which came when
They lived life too much and exploded.

So, we all star in the cosmic film,
Sometimes skipping some frames.

THE WORD FROM THE FUTURIANS

And it came to pass that God, Adam, Noah, Moses,
The Ten Commandments, the prophets, Jesus
And Our Lady of Fatima were of no real help
In reducing the sins of the horrible human nature
Invented by God Himself;

And so it came to pass even more, not less,
That all really appeared hopeless
For the higher mammal species known as Mankind
(women were secondary to God/Religion),
As homo sapiens had not
Even remained at the same level
But had gotten worse,
Some of this evil even being performed
In the direct name of God.

It also always comes to pass that God
Never gives up on his failures of creating
A temptation resistant species,
So, a new search is underway
For a teen-age virgin engaged to a saint
For GOD to impregnate with his daughter to be
(women are equal now),
Because, as ever, GOD still has no wife or girlfriend—
And is not even seeing anyone—
For how would a relationship
Between Mr. Right and Ms. Perfect ever work out?

However, so far, no teen-age virgins have been found
On the planet, much less engaged ones,
Princess Diana having been the last of the breed.

Furthermore, the virgin's husband-to-be
Must be saintly enough to let her fool around
With a Ghostly Superbeing for a one night stand,
Although, admittedly,
The Devil would probably be much more passionate.
So it came to pass that all this needed
To come to pass or humankind would be doomed.

Hail "Mary", full of Joe (hollow be his game),
Wherefore art thee? For the Lord, Art, in Heaven,
Harold be HIS name on Earth,
Wants to be with thee, blessed be thy womb,
For it is Heaven's Kingdom!

THE CREATION MUSEUM

The new Creation Museum in Kentucky
Features dinosaurs in their Ark,
Deemed small at the time (4000 years ago),
So that they would fit in the boat.
None were mentioned in the man-inspired Bible
Since their fossils were unknown at the time.
The oversight has now been corrected.

RAINBOW

Toward the end of a sunny day,
A storm came and washed away,
And the sunset clouds, being glad,
Held a party for the returning lad.

The sun then peeked, and soft shone
Into the mist of the departing squall,
Its light split into particolors lone,
Separating, each from the ALL—

A bouquet of colored rays
Swirled into sight,
And promised good weather
For the rest of the night.

The rainbow lit up the east,
As long we attended the feast
Of both the east and the west,
Till into darkness we descended blest.

The stars guided our homeward flight
By shining their jeweled lights
Of ruby, emerald, and sapphire
In living colors of blazing fire.

GRACIOUSLY WELCOMING
LADY LUCK BECOMING

He believed that luck would never fail—
So he ran like the wind through the jungle,
Surely knowing. He'd what he'd come for,
Now hopeful to find the help at the shore.

The relentless ones were not far behind,
That ill-fated menace of the bad kind.

Miss Fortune laughed, and said,
"No road could be too hard to tread
For we are fearless. To those, a boon—
For they ever seize the Opportune."

"I see you, Fairest Happening."

Just past a sharp turn, in the trees,
He suddenly dropped to his knees
And fired into his pursuers mean
As they came upon the scene,
Using all his ammo but for one round,
Then hurried on with nary a sound.

"I am wide aware," Miss Karma,
"Of this continuing Dharma—
That chance shines as my sun,
For, she, in turn, happens on everyone."

"Oh, say it is your lot, my friend and lover,"
She answered back, granting him cover.

Listening, he could hear ever more troops
Rushing through the night in groups,
About a half-mile back around the loops.
"I gratefully welcome thee,
Miss Lady Luck of Dice,
Though I may pay a late fee
For my pick up so precise."

Ms. Destiny Serendipity smiled, saying,

"The game is on; we are playing.
Let joy and innocence prevail;
Believe that luck will never fail."

He moved on, ever faster, cheating death,
A third wind becoming of her vaporous breath,
It blowing this DIA operative onward
To the shore ever toward.

He could hear the whirling chopper,
But now receding was its Doppler,
He thus grieving
Of its leaving.

"Am I much too late—still too far?
Shall I curse you all, destined stars?"

"No," said lovely dear Twist of Fate,
For you have one bullet left for chance,
Not to use to sleep or dream perchance."

But the chopper was rising high,
Well into the star-crossed sky.

"Shall to self I take this bullet
Now that the bus has left?"

"Oh, no," Miss Lucky Break encouraged,
"Do not be at all discouraged,
For you know it shall not be so
And what with it you now must do."

"Yes, perhaps it shall be so in some plight
Coinciding in a most kempt and hapful night."

He smiled and then knelt to ground,
And sent his last bright tracer round
Just ahead of the copter now departing,
His minor wounds yet sorely smarting.

"I bless you with all my lucky charms,
My good and well-fated man of arms."

The door-gunner noted the red tracer
And whence it came of the river vapors.
"Captain, turn back and take a look;
He awaits a fortuitous accidental fluke."

"I am an uncursed, non-jinxed agent man.
Let my joyous innocence prevail again."

He jumped into the rescue's hovering haven,
Directing the door-gunner's firings, wavin'.

"Fare thee well, my nightly knight"
Dame Fortune wished upon his sight.
"You recognized me even in the dark."

"Oh, My Angel, Passiona, lovely lark,
I might have known it was you
That would ever see me through."

HERE

No, I did not just disappear—
I am just completely soaked in her qualities.

The drop has become the ocean—
Now I drink from her spring of eternal youth.

Do I feel some memory of elsewhere?
Do I dare to look into the setting sun?
No, I'll pretend that it's coming up.
It shines through me, illuminating me.

I am re-energized. I am glowing bright.
I am becoming a supernova.

The world crashes, out there,
But the flowers grow, in here.
For, I am the garden.

JENFINITY

Jen wends to her den in her home,
To pen the yen of her endless tome,
The treatise of limitless immensity
On the subject of boundless infinity,

Utilizing her ken of the ever-when so vast
As to cover the present, now, and past
Of the chicken and the egg, zen,

The multitudes being the mother hen
Of all mysteries born now and then,
Which are a google to the power of ten.

The abundant profusion of everything
Sent her to us, and that same shebang thing
Has now taken her away, leaving nothing.

The whole ball of wax is unwound,
The whole nine yards lying on the ground;
Kit and her caboodle are not here—
Thank you, and good fortune, dear.

LOVE STORY

A man falls in love through his eyes,
A woman through her ears;

Later it reverses...

A woman notes everything to be done,
But the man does not hear the seeing one.

But there is still hope...

As in the marriage
Of the blind lady
To the deaf man.

—INTO THE LAND OF ALL THE GODS—

Into the realm of supernatural figmentations
I drifted off, within my newest imagination.
To interview all the living Gods there;
Some who've left and some yet ruling everywhere.

Notions of 'God' are of the wide purview
Of the inquiring mind confined, their 'why'—
That wide expanse of fables, faith, hoaxes,
Lies, imaginations, fictions, guesses,

Foggy notions, concoctions, phantasms,
Fantasies, falsehoods, conceptions,
Decrees, fiats, misrepresentations,
Dead ideas, magic, proclamations,
Wild tales, anecdotes, revelations,
Untruths, revelations, hearsay, scrap heaps,
Yarns, and fish stories stated as beliefs
In that unseeable supernatural station
Through faith's without knowledge ration;

They are all figmentations of the imagination.

Strewn about this great panoramic realm
Of the ONE possibly conceivable at the helm
Are all of the unknowable fabrications
Often Dreamt up via exaggerations
By the human race of mammal sapiens.

The realm of such Pronouncements has come to be
Superposed at the furthest edge of Reality,
Poised by the scope of some wishful thinking,
By all those dreaming and wild supposing,
Who wish for such legends to be ever
Actualized and realized; however,
These ideas have never ever made it
Into our observable realistic habitat
In any way, they but remaining in the minds, joint,
Of the God-beholders—even as varying viewpoints.

Without so much much as a word to say,
I passed those to whom most no longer pray,
Nor believe in, but once did, namely,
Those of the graveyard tombstones now so unholy:

Astrology—the God of the Stars that plod,
Eternally blazed and marbled in the sod,
The monuments of Diana the Moon God,
Apollo the Sun God , Baal, Zeus, Wotan,
Aphrodite, Thor, Mithras, Isis, Amon,
Poseidon, the Druid Gods, and on and on, anon.

I ever hurried past the ledgering
Of those older Mythologies preceding
The formation of the Old Testament story—
Those ancient superstitions whose very
And various olden amalgamations
Came forth to form it whole for our salvation.

I paused at that Old Testament maligned,
To mark the old but lingering lines
Of the 'knowing' of more invisibles,
The beliefs in imagined angelic creatures...

There were Angels standing, frozen in stone,
Over the timeworn memorials' poems,
And atop the crumbling gateposts,
They being the livelier and near-living ghosts
Of the representations of the three spheres

Of the heavenly host: the demigod-near
Seraphims, Cherubims, Ophanims,
Thrones, Principalities, Dominions,
Powers, Archangels, Angels, and, those final,
And perhaps the most useful—the Guardian Angels
That are said to protect children from falls.

There, Amaranth, its dead red leaves never
Ever fading on this Earth unto forever,
Gave some color around the graveyard pallor
And to the dateless headstones' squalor.

There's a garish purplish maroon view, on high,
Of streaking lights of an electromagnetic sky,
Heretofore never imagined by my self;
I strolled on and into the vale itself.

Every-thing, every order happens for a reason?
Yes, for the most part, for most seasons,
But not for the bottommost cause the first,
For there was nothing before it to reason it forth.

The Intelligent Designer

I approached a semitransparent
Theistic Embellishment, rather well lit,
Who was holding out an eyeball—a shove
Of His hand for me to take note of.

"Who might you be?" He proclaimed,
"For I am the God of Intelligent Design,
The One who was made by the signs discerned
When the creationists noted them all unlearned.

Lo, They saw inexplicable complexity in Nature,
And, thus, they leapt and promulgated that Nature
Must have a Grand Designer of its mechanical dance,
For how could life have come about by chance?"

I replied, "You're right about chance's stance,
But wrong about chance, too, for little greatness,
If any at all, comes about by mere chance,
Especially as some giant leap in one bound
Up the sheer cliff-side of Mt. Improbable—

To find on its top a great complexity
Of something like the eye that You show me;
However, it is actually an error to suppose
That Chance is the scientific alternative
To Intelligent Design, for that's quite negative.

Natural Selection is the means of the design,
For it, unlike a one-shot chance, being not in kind,

Is a cumulative effect that ever winds
And slowly and so gently climbs
Around the mountain's other side, behind the sight,
To eventually arrive at the great height
Of complexity—from which we can then view
The beautiful sights through our eye anew."

"But the widespread Watchtower Zines
Always pronounce that the biological Designs
Were created by Me instead of by chance!

Just look at these eyeballs—take a glance—
And the optic system hanging behind them!
How could that come about by chance, these gems?"

"You, like your followers, may listen,
But You do not hear, writing with untruth's pen.
IDers deceive by this wrong approach,
Whether they mean to or not; I give reproach.

Chance is not the opposite of Nature's design;
Evolution of the Species through the graduality
Of Natural Selection is the path to complexity;
Your ploy falls as flat as an imaginary line.

A flatworm has but an optical system's spark
That can only sense but light and dark;
Thus, it sees no image, not even a part;

Whereas, Nautilus has a 'pinhole camera' eye
About as good as half a human eye
That sees but very blurry shapes;
Thus they are examples of intermediate stages.

'Rome' can not be built in a day by chance;
Chance is not a likely designer at all!
Really now, could a 747 ever be
Assembled by a hurricane blowing free
Through Boeing's warehouse of all the parts?
Now is this the sum of Your conversational art?"

"No, Austin—it's quite unlikely—'tis just to confuse,
And that's why we always so misleadingly use
This 747 argument as the contrast to ID...

So, then, Austie, chance and Intelligent Design
Are not the two candidate solutions we'll find
To the riddle posed by the improbable?
It's not like a jackpot or nothing at all?"

"'God', Your ID ideas persist, as repetition,
But, again, chance, for one, is not a solution
To the highly improbable situated Nature,
And no sane anti-creationist, for sure,
Ever said that it was; your tale is impure.

Intelligent Design, is neither, a solution—
Because it raises a much bigger question
Than it solves, as You will soon see, in a lesson."

"Well, I'll be darned," replied the Designer.
"Natural selection *is* a good answer;
It is a very long and summative process,
One which breaks up the problem's mess
Of improbability into smaller pieces, less,
Each of which is only slightly improbable,
But not prohibitively so, thus it's reasonable
As the product of all the little steps, of which,
Would be far beyond the reach of chance—it's rich!

The creationists have been looking askance,
Seeing only the end product, perchance,
Thinking of it as a single event of chance,
Never even understanding
The great power of accumulation.

Such they didn't know much else—their fall,
Not having any other natural ideas at all,
So, they outright claimed that ID did it, as the Tree
That can magically grow the All, namely Me."
"So, 'God', You have now seen the light
Of the accumulative power's might;

This is the elegance of Evolution's 'sight'."

"Yes, but what is to become of Me, the Person,
For I only 'exist' through their speculation.
In fact, the improbability of Me is so High,
And so much more so from where I lie so 'sure',
Compared to that of 'simple' Nature,
That My own origin..."

"...is a near-infinitely LARGER dilemma, Mate,
For the creationists—the problem that they love to hate;
That being that You, therefore, can only be explained
By another, Higher Intelligent Designer claimed.

Far from terminating the endless regress,
They've aggravated it with a vengeance
That is way beyond repair or redress—
As beyond could ever be yonder of! Out west!"

With that, the poor Guy faded toward oblivion,
Which, remarkably, which was the very location
I was visiting, but, hence he soon reappeared,
Although in another guise, but quite well attired:

[God created Adam, then Eve, of Adam's rib,
Both fully formed, imbued with God's knowledge
And memories of times that never were,
Such as childhood.

They believed a shifty talking snake,
Ate the verboten fruit,
And were cast out, to fend for themselves,
God being quite surprised at their sin...]

The God of Irreducible Complexity

"Hello, Austino, it's time for more perplexity;
For I am now the God of Irreducible Complexity."

"That you are, being the unmade All,
And so it shall become your downfall."

"Eh? I'm never to be at all?"

"Your believers have given You some fine new clothes:
But, Intelligent Design is falsely based, God knows,
On Irreducible Complexity—
So I still recognize You as the God of ID."

"That I am is what I really am now."

"Well, Darwin said long ago that his theory
Would break down if Irreducible Complexity
Were shown to be true, and, yet,
No proposal has ever stood up to the analysis."

"Still, here I am, Mr. A, alive merely by possibility,
Myself indeed quite complex, even irreducibly,
For I am the be all and end all—the Prime Maker,
And so I keep tabs on every form and splinter
Of the Universe, planning its every constituent
That I designed. So, then, simple I am NOT.

Yes, man, I am an extremely complicated System,
Yet I have no parts, for then My parts that stemmed
Would be even more fundamental than Me!"

"Yes, 'God', if You existed you would surely be
Very very very complex,
Irreducibly so..."

"...So..."
"...So, by the Creationist Theory, such as it must be,
You cannot be explained except by a larger ID."

"I'm falling..."

"...into the hole that they dug for you."

The God of the Gaps

Yet another Theity appeared, out of the mist.
"I am the God of the Gaps, of all those missed.

I Myself personally fill in all the gaps withstanding
In the present-day knowledge of non understanding,
Albeit a very large and unwarranted assumption,
But I surely do fill them all in—via the fiat lent
To Me by the creationist's fine endorsement."

"These gaps shrink as science advances anew."

"And so there is less and less for Me to do."

"What worries me is not so much that You
May be eventually laid off, having nothing to do,
But that those of Religion think it is a virtue
To be satisfied with not understanding a quandary;
Enigmas drive scientists on—they exult in mystery."

"True, My believers exult in mystery
Remaining as mystery,
And so they go no further,
But, it keeps Me from being history!
They worship all these evolutionary gaps as being Me."

"With no justification?"

"We have a 'get out of jail free' card—a vocation;
It's an immunity to the rigorous proofs of science;
We just claim by the 'say so'.
All must respect that stance."

"You lead a charmed life, then, one with no faults,
But You seek ignorance in order
To claim victory by default,
As a weed thriving in the gaps of science's fertile fields.
Scientists rejoice in (temporary) uncertain yields,
Whereas You halt all inquiry."

"I remain as a mystery."

"You're the same God of Intelligent Design assumed—
Now known by a much more desperate nom de plume."

"I repeat that I intervene to fill the evolutionary gap.

I even alter DNA."
(We could check the evidence for that.)

"We researchers fill the gaps in the fossil record."
"Then there are twice as many gaps. Absurd."

"I'd laugh, but I know You're not joking."

"No joke. Try what we've been smoking.
Lack of 100% complete documentation
Of Evolution means that I aid its motion."

"'God', That is not a good default stance."

"It's an unknown happenstance."

"So, do we let criminals go
Because we don't have a video
Of their every intermediate foot step
To and from the lawless event?"

"No, of course not, but we now have great worry
About our precariously perched gappy theory.
Also, you made a typo—it's a <u>God</u> default stance,
Certified by nothing more than proclamation
Of Our Bull of Decree covering all instantiation."

"An edict, huh."

"Why not, duh."

"It was also once avowed that an Evil Spirit,
One that You Yourself allowed to exist,
Produced physical illnesses, on us weighing,
But, thank God—just an old saying—
That scientists persevered, and still do,
Such as finding out about the immune system's zoo—
Our defense against the non evil spirits
Of germs, viruses, and bacterial fits."

"Yes, agreed; that claim was dead wrong; take pills,
But evil spirits still cause the nonphysical mental ills

That are called sins and bad thoughts,
Even crimes of wills."

"Still trying to halt scientific inquiry,
I see, for the burning.
Mental lapsing 'sins'
Stem from upbringing, wrong learning,
And/or low serotonin and such imbalances, needing cures,
Not to mention the many differences in cultures,
Such as other religions being a problem of stability,
For people think this undermines
Their own belief's credibility."

"Okay, I give up for now, AustinTorn. Be.
Go on with your work, with My blessing,
To discover important truths about reality,
But some fossils are evidently missing!"

"Only a tiny fraction of corpses fossilize;
However, not even a single fossil guy
Has shown up in the wrong geological stratum;
How's that for absolutely no erratum?"

"Well... it's sad for Me, but true.
I'd still love to find wrong a few,
Like a fossil rabbit in the Precambrian.
I'd have planted one there if I exited then."

"Dream on. Lazy reasoning is all that's behind
These declarations of the irreducible complexity kind."

"Yes, but all this ignorance, for sure,
Of the possible steps of Nature
Has kept Me forever alive,
Allowing Me to ever thrive."
"And has just as soon forgotten You, in truth,
But for those sustaining your being without proof."

"Wait, what about an arch of bricks?
(I'll try to use this one as a trick.)
Pull one away and the arch falls apart;
It cannot survive the subtraction of a part,

So, how, then, was it built in the first place?
With this insight I can win the human race."

"By scaffolding, the same as seen in Evolution."

"I was afraid that would be the solution."

With that, the holely God of the Gaps separated
And nearly evaporated
To become a discontinuity Himself,
But the creationists gave Him help
By Holding Him together
With their last ditch effort.

(Yes, 'gapping' still goes on, it seems.
When the argument first gathered steam,
There were but a few transitional forms known,
Although good ones, enough for the idea to own:
One being the bridge to vertebrates
And another the bridge to flying creatures.

But there are many more now, a wide range,
So, then, it is the data that has changed.
These 'gap' arguments were already down
To the faint hope that scientists, as clowns,
Wouldn't find any more natural explanations;
But the finds were the most inevitable situations.

Creationists yet remain at the pointward
Of not being able to 'push forward',
So, all that's left to is push backward,
Albeit at the firmly established fact words
Of evolution. Even the Pope concedes this,
But tries to salvage the faith, and solve,
By saying that the mind was not at all involved.)

In the darkness I alit from the Wiz,
And tried to make sense of this world of His.
Now I've found the answer to life's dark quiz:
One must live this life by what light there is.

The Deity

Another God appeared, a mere Deity,
(meaning no intervention, so He's not a Theity),
And thusly said, "Forget the Theity solution.
I am the Smart God that seeded Evolution.

It was I that set the whole universal notion
And all of life's evolution into motion;
That was My elegant and foreseeing way
Of creating the kind of life that would stay."

"I thought You were all powerful;
Why not just make 20-40 million species
All fully formed, as immutable as Thee,
Along with their usable natural habitats,
For this is how most Gods would do it?
What energy loss could that be to You?
Your infinity could all this in an instant do."

"I'm not so Great, plus, since Evolution is too stable
For some creationists to scoff at, as a fable,
They have assigned the job to Me, the Creator,
As all of Nature's natural Instigator,
Because, they must take retreat from the first ID God
Who zooms souls into humans at birth—it's so odd.
So, now I am not a Theity any more of proof
And thus I must ever remain aloof.

Of course now I have very little to do
And so I am not much needed, true,
For I can't even muddle with their lives;
They are all stuck now with their wives.

I might really just as well retire,
For I am superfluous and tired."

"Well, You're still kind of close to our Universe,
Not completely outside it, maybe, the place the worst,
As I suppose your successor will have to be placed,
Absolutely, totally invisible to the human race.

At least You made some basic primordial substance
And foresaw the billion years of combinatorial chance,
Predicting every turn,
Or at least knowing that something neat
Might probably come out of it,
Which was still quite a feat."

"Thank you, but it was nothing."

"On the contrary—
I say verily—
You're <u>the</u> Super Scientist,
An Engineer Par Excellance—
The Ultimate Inventor of All Time—
Much better than than the old God of ID."

"Yes, I am a Scientist, making all that's real—
I HAD to be, but it was really no big deal."
"You're too modest."

"It was just some little quarks
And some electrons that I sparked
And some forces that arose
As reality was composed."

"But look what became of its simplicity—
Through its stages to astounding complexity
Over billions of years of circumstances;
We've traced the composites back to simple substances."

"Well, um, it did really take that long for My intention,
By some coincidence the same as that for evolution;
however, I guess I'm just as surprised as you, frown,
That when some examine substance and get down
To these simple subatomic levels of unadorned things,
That they then take a giant leap back, of all things,
To the composite complexity of Me, the Ultimate.

Isn't complexity a much higher product
Of combination upon combination,
And thus not lower than simplicity itself?"

"Yes, it would seem so; that's a near empty shelf.
Then I suppose You're some Great Alien Scientist, odd,
Highly evolved from somewhere else, but not really God."

"True, and you, Austin, as a scientist,
Should seek what underlies the all,
Not some Great Complexity who oversees it,
For that's for what the theory calls."

"Wise thoughts."
"The best that can't be bought."

Well, whatever on the alien thing of it,
But the creationists are not keen on scientists,
For scientists regard the honest seeking after truth
As as supreme virtue beyond all reproof.

If they ever found out..."

"Yikes, they know not what they have made Me.
As a Scientist Myself, I truly value honesty
And skepticism over the dishonestly faked beliefs,
Those that only seem to bring Rolaid's relief."
"The Founding Fathers of America liked You,
Although some of them, as Thomas Jefferson, too,
Were outright non theists, many seeing You as a Deity
Who just started things up,
Never interfering with reality."

"Funny how President Bush's America sings,
Straying so oppositely from its humble beginnings."

"Not to mention that some the world's peoples, really,
Are squandering their precious time
Worshiping a Theity,
And sacrificing to Him, begging, fighting,
And dying for Him,
Even threatening the world with its destruction."

"What a waste."

"Are you real?"

"No, I am but a figment of imagination, see,
But some really do like harmless old Me."

"So, what's really fundamental?"
"The real fundamentals, just below
What you now call 'fundamentality',
Have always existed—the quantum reality."

"There's perhaps no time of 'forever'
At that level for Your 'always' ever."

"True, they just are, and had to be—the possible,
For a state of absolute nothing is indeed impossible."

[God therefore became very angry
At His own lack of foresight,
A base emotion of His own invention.
The harshest punishments of all time
Led to all of the gruesome events
Of existence portrayed in the Old Testament,
Rather a waste,
For man's past and future sins were soon redeemed
In the New Testament by God's Son, Jesus,
Conceived of a virgin.]

The God of the Agnostics

I came next upon a God sitting on a high fence
And waved to Him, saying
"Come down and talk the whence."

"I can't; I am stuck here, but Salutations to you.
I am the God of Agnosticism, one neither false nor true.
None of the agnostics know if I exist or not,
So here I must stay put a lot,
Along with the Tooth Fairy,
Santa Claus, and the Easter Bunny,
Just in case we all might exist or not,
As a quadzillion-to-one shot."
"Why can't agnostics make up their minds?"

"My followers cannot even make or see
Probability judgments about the question of Me.
This is the limitation of agnosticism,
Perhaps the error of no consideration
Of the likelihood of that for which evidence seeable
Is not even the least bit conceivable."

"It is a fallacy; what I call the poverty of agnosticism,
Because, although being agnostic is reasonable criticism
For some things, such as whether life exists elsewhere,
It is not appropriate for those things undoable
for which the idea of evidence is not even applicable;
However, actually, we <u>can</u> actually still talk
About the probability of the event,
While even going for a walk.

The true fallacy, however,
Is that the existence ever
And the nonexistence of You never
Are not even on an even footing to begin with.
The two are not at all equiprobable cases.

The burden of proof lies with the believers,
For anything that we can conceive of
Can be claimed to exist, as that we love,
Such as ghosts, spirits, and such forth.
Are we then to straddle a fence that has no worth?"

"And, never seen. So, then, at the end of the day,
Probability creeps into the beliefs of the agnostic way,
For in practice they end up in the lurch,
Not going 'half the time' to Church,
But mostly deciding not to go at all."

"Yes, they still decide that which is 'undecidable',
For the fence is very uncomfortable
And so. then, the superposition
Decoheres into the inclination
Of non belief—until right here
The Extraordinary's evidence appears."
He came down off the fence,

For he couldn't exist and not exist at the same time.

I continued on through the undulating hills.

(We can refer to the fence sitters as non theists,
In order to get away from labels like 'agnostic'
Which might imply that the probability of thinking
God, or not, is on some kind of equal footing;
Plus, that the fence sitters don't really stay
On the uncomfortable fence but usually...

Go one way or the other way
In life's practice of the everyday,
Although some might go to church
On alternating Sundays.

In between,
Perhaps they go on wild picnics with their sweetie
And drink wine and do all that 'bad' stuff
That we can't say here
While waiting for some extraordinary evidence to appear.

I will soon have a talk with old Jehovah Yahweh's Thee.
He's not so terrible as many have made Him up to be,
But then again He's not so great either—He's quite off,
Just another poor middle manager
Caught up in the layoffs.

I spoke to the Deity—the God who doesn't ever interfere
In the running of the universe.
The Pope doesn't know it here,
But a Deity is what he's leaning toward when he says, then,
That evolution is acceptable now
For Catholics to believe in (no mind).
The Deity Guy was actually
Kind of a great scientist.

I already met with the Creationist's ID God(s),
Who, while still a Designer,
Is, well, not so cool at all, either,
For He gets back to what the Fundamentalists believe,

And neither, they would say, did evolution happen,
Or, if it did ever function,
God constantly stepped in to rectify its direction.

I haven't really begun
To scratch the surface of all the Gods,
Though, for so many lie now beneath the sod.
I'm only interested
In the person-type Gods of monotheism,
And I'm hardly even
Getting through those variant theisms
That fight amongst themselves over Jesus' divinity,
Or if there is a Hell,
Or a Devil and some Angels about thee,
And over so many more other major differences, totally.

Then there are the multiple Gods, now up in the millions,
And also the many gods-who-are-not-persons,
Plus the TAO,
The Consciousness and some way-out Ones.

There are also hundreds of long gone, 'sure thing' Gods,
Which I needn't get into, except to wonder, and say:
Is that how the future will look at our Gods of today.

I can also skip the many weird offshoots that persist,
Like those saying that the self is not allowed to exist,
Even calling it the 'ego' to make it seem so much worse;
I don't have time for these and other cult-level verse.)

The God of the Old Testament

Of all my rotten luck, The God of the Old Testament
Appeared and proclaimed, "I am Yahweh, never absent,
For, those schooled from infancy in My strange ways
Have become desensitized to MY horrific side
And so they continue to keep Me very much alive
Through their thoughts; so, Fire away at Me;
I no longer bite that hard, you see."

"You're too easy of a target to attack for free—

So it would be rather unfair of me."

"True, and I won't deny it—
It's all there in the Testament.
I was the most unpleasant character
That anyone ever made up in literary fiction.

I was revealed to be jealous and proud of it,
Petty, unjust, controlling, vindictive,
An ethic cleanser, genocidal, infanticidal,
Filicidal, pestilential, megalomaniacal,
Homophobic, misogynistic, sadomasochistic,
And much more, and a Bully—who gave it
Free will only if it matched My own Will."

"Peace be with you. How about the New Testament
To replace and hide Your scent,
As many religions have already done through Jesus sent?"

"Yes, that Testament is quite opposite in tone,
But I am still the Father of Jesus sown,
So, the problem of Me can never really go away.
I am what I was, still here unto the present day."

"Well, so long. You're the worst role model, yet,
That human mammals have ever dreamed up.
Who would imitate, emulate, or follow You as a 'leader'?"

"Well, My followers are those numerous slaves
Who excuse my mysterious [insane] ways,
Along with my exclusive desert tribe."

"Well, You're the Boss, and, anyway,
Who ever said that a God had to be perfect and good?"

"Everyone that I told—and those who thought I should."

"Oh well, never mind; whatever pleases.
So, um, Joseph was not the biological father of Jesus?"

"No, I was."
"So Jesus didn't really descend from David?"

"That was on his mother's side."

"Well, my ancestors descended from the trees.
Hey, why don't Catholics get the 72 virgins
That Islam gives for martyrdom for their sins?"

"I told each religious faith a different story."

"You also gave a bible half-different to the Mormon founder,
Joseph Smith, finely engraved
On golden plates he discovered?"

"Sure. I thought at the time 'why not'."

"You had Islam add different things to their Koran, as well?"

"Yes of the many more ways to avoid Hell."

"And You told only the Catholics
That there were umpteen levels of angels
And that bread was your body
And that wine was your blood?"

"Yep, I told just them and a few other selves,
But they made up the Saints themselves."

"And You presented differing visions to the Lutherans,
The Episcopals, and the Jewish,
And to many other also-rans?"

"Pretty much, except that a King of England
Founded the Episcopals—the Anglicans, of course,
Since his own religion wouldn't give him a divorce."

"And you killed everyone but Noah
And his family in the Great Flood, wet,
Even young children and their pets?"

"Sure, again, why not? Life is cheap.
However, My creation of the rainbow
Says that I'll never be so cruel again.
What can I say—I goofed. My sin."

"But You are infallible, and even omniscient
And so You know all of the future meant."

"My omnipotence of changing my mind got in the way."
"But your omniscience knew you would... one day."

"Yeah, I know—it's a paradox; oh, the strife.
And I can still technically end all life
By means other than a flood."

"You burned people in Hell, not saved,
When they didn't follow the unfree will that you gave?"

"Yes, because I was not a loving God."

"Who made You, God?"

"No problem—either I was Eternal or I made Myself be"

"This is remarkably the same, but for Thee,
As the Universal ingredients would be."

"Then who would need me—wait,
I don't want the answer told."

"Is the Earth only about 4000 years old?"

"Of course not, but I may have let that slip to some
To tease their intelligence apart from being dumb."

"Do you mind-read the thoughts of every human,
Using all of your acumen,
And write the earthly script for each event,
Being so omnipresent?"

"I tried that, at first, but it didn't work for Me
To put my finger on every atom that be
And micromanage its doings for all of thee."

"That's called 'God's Will', by some, even now.
What went wrong? Was it the where and how?"

"It disrupted the atoms' normal and natural movements."

"And that's what caused the storms unfocused,
The lightning bolts, and the plagues of locusts?"
"Yes, so I stopped making such a mess of things."

"So, the prayers of six million Jews, pleaded,
In the holocaust, went all unheeded?"

"Yes, plus I have better things to do, in time,
My sooth,
Than look after some old experiment of Mine
From my misspent youth."

"Did you really make Adam and Eve
And all of Earth and Nature, as we believe?"

"Yes, I made Nature, including the humans,
In My image."

"It shows in their rage."

"Thank you."

"God, it's ID deja-vu all over again—
I really have to move on."

"No, wait. I like your questions.
I'm mellower now, this being My new direction.
Not as many strictly admit to Me anymore."

"How come so many of the gospels were omitted
From the New Catholic Testament,
Like those of Thomas, Peter, Nicodemus,
Philip, Bartholomew, and more,
As well as whole books kept from us,
Although You told some other religions to keep them,
Such as the Book of Revelations?"
"Those gospels were embarrassing and wild;

They told about My Son doing magic tricks
And practical jokes on people when He was a child."

"Oh, we never heard much about his youth.
And didn't You send the Mormans proof
That Jesus spent an early era
In what was to become America?"

"Probably."

"What about the trillions of galaxies in the sky?"

"They're just for show and scenery on high."

"Where's all your rantings and ravings
That I've heard about?"

"I now take Prozac for My mood swings and bouts."
"You don't really exist, do You, as mental,
For how could You have an emotional system—
A composite—and still be absolute and fundamental?"

"No, I don't exist, for how could I since I am so horrible?
Human mammals made all of Me up
As a very bad example,
As it turned out, from their many fears
In the childhood of their species' years.
Unfortunately, it caught on to their children's ears."

"So, yet You still subsist
In this indefinite locus of wishes?"

"Yes, sort of. I am sustained here since many children
Have learned to obey and listen
To what is/was told to them,
For this obeying was an evolutionarily useful thing,
As many of their obedience
Resulted from warnings of things
That were truly dangerous, and so the children grew up
To indoctrinate their own children
In all the 'knowledge'."
"We'll have to offer more reason

To those so indoctrinated.
Now farewell to You, the impersonated."

"See you. Pay no attention to Me as certain,
But to all those blinded by the curtain."

He soon dozed off into never land.

In the beginning...

God obtained some material somewhere
And thus created the Heavens and the Earth.
Eve then cost Adam rib, then an arm and a leg.

Cain killed Abel and so we are all Cain's children.
So, our ancestors descended not from the trees.

Noah married Joan of Arc and took her
And all their pets on a world cruise, noting
The rest of the human race as dead and drowned.

God played a joke on Abraham,
Whose kind had often made burnt offerings when
Popping the corn or overcooking the Lamb of God.

Moses then tied his ass to a tree
And wandered away to cleanse the tribes.

The ancient Egyptians fleeced the electrolytes.

God made spiritual love to a teen-age virgin,
And Jesus was born, died, and was born-again
On Easter (Let us Raise the Lord!),

But not before Jesus had made water into wine,
Perhaps encouraging alcoholism.

Mass was served by the altered boys
And even odd girls, all preyed upon.
Lent soon became fast-food only time and
So Fat Tuesday was invented to tide one over.

Thank God! Sleep be with you. A-choo.

The Gods Meet Each Other

I next encountered all the individual Ones,
The specialized Gods of all the Religions.

They didn't get along at all, not even for an instant,
For all they had in common was their intolerance
Of the others' greatly erroneous and unjustifiable beliefs
That clashed with their own, for tolerance as a relief
Was truly NOT an attitude the jealous Gods endorsed.

The followers of each God thought that their own
Irrational embrace of myth trumped the others' known,
And so this led to many of the religious groans.

I watched the Gods battling for a while, steadfast,
In the present, as well as in the distant past,
Their followers' beliefs scripting the actions
Of conflicts that led to dying for untestable propositions
About where everyone came from and was going to:

Metaphysical Martyrdoms
Conflicted with the Divine Book of Revelations.

Deuteronomy 13:7-11
Stoned those disbelieving in Yahweh,
Killing them, while the Koran eliminated many infidels.

India and Pakistan, different countries domiciled,
Because the beliefs of Islam could not be reconciled
With those of Hinduism, poised themselves at the brink
Of nuclear war, merely because they disagreed, rife,
Over some supernatural 'facts' concerning the afterlife.

Karmas ran over Dogmas.
Musharraf suspended Pakistan's constitution,
To stamp out the growing Islamic militant coalition.

Palestine's Jews and Muslims scuffled on;
Balkan Orthodox Serbians dueled
With the Catholic Croatians,
As well as with the Bosnian/Albanian Muslims;
Northern Ireland Protestants warred with the Catholics;

Sudan Muslims discorded with the Christians;
Sri Lankas's Sinhalese Buddhists
Went against the Tamil Hindus;
Caucasus Orthodox Russians and Chechen Muslims
Exterminated each other and their kin;

Iraq's Sunnis and Shias massacred each other
For some very slight dogmatic differences.

I interrupted their skirmishing and said in haste,
"What about tolerance and respect for the other faiths?"

They all answered at once and said, in unison's beef,
"That's just political talk. If we tolerated other beliefs,
That would be akin to recognizing them readily
As having some credibility,
Which they certainly do not."

We are saved and they are all doomed, in peril;
We can't have them exerting influence in the world."

"So," I said, trying to make some small talk,
"I've heard that You've each written a book
That makes an exclusive claim as to its infallibility.
Congratulations to each of You on being published.
All have made the bestseller list.
However, I have respectfully shelved all of them
Next to the *Egyptian Book of the Dead*
And Ovid's *Metamorphoses*
In the contradictory book and Bible section.

Hey, how about getting modern and making a film?
I know that a book was a great thing way back,
But a moving picture is worth 10,000 still pictures
Which are in turn each worth a thousand words."

"Indeed, we will each be divinely inspiring a movie
That will soon be playing in a theater near you."

"Wait, Guys, I take it back," I said with alarm,
"Are not all your children doing enough harm
By fighting over your books and morality plays?
Will people now die for another media—the movies?"

They ignored me and fought on, with their kind,
Unable to see but through their own 'right' minds,
(Doing the opposite of their teachings of love)
Which they were especially and paradoxically out of.

Unfortunately, they now represented the largest threat
That human kind has ever imposed against itself—
All due to differences regarding some very improbable
And differing notions about the nature of the universe.

We Are It

Appearance and motion wholly create
Being and time in the arena of space;
We're the complex composites from simple verse,
The ultimate, so far, in the universe.

I noted the Land of Evil Demons,

Although sometimes
It was hard to tell which
Was which or not witch.

I also bypassed the numerous Gods
Of the instant Cults
That had always gained so many followers
And bad results.

The God of the Religious Moderates

I next encountered the God of the Religious Moderates,
Whose numbers had been swelling lately, at any rate,
But they had seemed to get stuck in that middle state.

The God of Moderates said to me in soft oration,
"Greetings. All things in moderation."

"I bet that You derive from secular knowledge
Combined with religious ignorance."

"Well, yes, modernity has allowed some dust to settle
On the very old unchangeables that do nettle,
And so now people pick and choose,
Invent, or ignore the Dogma's ruse."

"Dogma is indeed an unchangeable definition—
It does not admit of progress, by its very definition."

"True, but I am still their God, of course,
As they have abandoned the wingèd horse,
Virgin births, sexual prohibitions, the value of life—
And they even have some doubts about the afterlife."

"They betray both faith and reason."

"That they do in this new season."

The God of Nature

Lastly, I met the God of Einstein—Spinoza's God.

"I am the so-called God of Nature,
Being am one and the same with it—no different;
Although, that which has no difference
Is really not any different.

Anyway, at least this is how the people awed
By Nature's intricacy and beauty refer to Me.

I am only here in this nebulous vicinity
Because I don't actually exist with certainty,
But seem to some to be tautological with Nature,
Always existent and beautific."

"It's OK, don't worry about it."

"Thank you, and welcome to reality."

"You mean I'm back?"

"Well, at least you have one foot in it through,
Just as I seem to do."

"I'm going, but why did humans
Invent the theistic and deistic Gods?"

"Man created them in his image's inward glance
Because he was and is terrified of his insignificance,
As well as from a fear
Of losing the beauty of his life's instance."

"So man just proudly declared
That he was of Special Creation."

"Yes."

"Farewell and thank You for Your insight."

He called after me.
"Enjoy reality—it's really a place that's better.
There's nothing more beyond it. ALL comes from matter.
You're electrochemical creatures—
As organic and natural as anything else in Nature.

Consider this knowledge as the ultimate humility,
If you will.
Live life, love it—while you can,
During your lucky incarnation
From the evolving composites
Of the last 13.75 billions years.
You are here. You have arrived."

The Nature of Pantheism

Panthea, the greatest God there never was...
How to explain? She does what nature does.
As a rose is still a rose by any other name,
Then so is a universe a universe the same.

Down to Earth

As I rejoined Actuality, I felt its waves and seas
Of brightness and color joyfully washing over me.

Getting back to my existence and its stresses,
I ignored some knocking Jehovah Witnesses
Then made nine golden tablets,
And reported my findings on ToeQuest,
And then went breathing, seeing and hearing,
And otherwise sensing all that was knowable as reality.

The Problems of Traditional Religion

The Christian concept of reward and punishment
Handed out by an omnipotent, omniscient God,
Is derivative of the family experience—
The child and parent—a conception of our world.

The News

I picked up some newspapers and magazines:

A suicide bomber blew up a bus and himself as well,
Sending many of the unbelievers straight to Hell,
While assuring himself and 72 friends a place
In Heaven, a double blessing from his Faith.

His family, relatives, and friends gathered, soon,
To celebrate their wonderful good fortune.
The bomber's death was especially lauded as wise

Because he had proceeded directly to Paradise,
Bypassing the possibly troublesome way
Of the litigation of Judgment Day.

Fighting continued in Kashmir
Due to some perceived insults to Muhammad.

A man was released in Northern Ireland
After claiming to be a Protestant atheist.

A child of Christian Scientists died
Due to the religious refusal of antibiotics.

Extremists sought nuclear formulas and parts to reduce
The peril of the unbelievers in the world,
Those whose ways are not sanctioned by Allah.

Pope authorizes millions to reach
Children sexually abused by priests.

The recently discovered Gospel of Judas
Suggests he wasn't really such as bad-ass.

Some nuclear facilities no longer exist in Syria,
About whose disappearance both Syria and Israel
Seem to know nothing about.

Battles rage on over differences in some holy books.

Iran promises to destroy Israel.

President Bush led off his latest speech with
'In God we trust."

And in a more than 2000 year-old newspaper:

The Emperor led off his latest speech with
'In Zeus we trust'.

And, finally, in a future newspaper:

Religious extremists detonate atomic bomb

In Washington, DC;
Nuclear retaliation destroys
Twelve highly populated middle-eastern cities.
World greatly stunned, begins to widely read
The End of Faith', 'The God Delusion',
And 'god is Not Great."

THE ONE WHO HAS IT ALL

Of what fires did God's heart beam so bright?
Who or what could this all-seeing one make?
From what strength was His power made?
Wherefrom His great talent bestowed for free?

Oh what the luck that He was born so great,
Being Himself the one and only King?
Oh so gifted that all things came naturally,
All known so effortlessly throughout Eternity!

What a fortune He was given, living in Heaven,
For He there had it made, living in the shade.
Oh, He must ever so thankful be, for being He,
Enjoying all His riches throughout Eternity.

SIX TONS OF GOLD

The Overload of the 'master race' being born,
Was ever attracted to the amber room adorned,
And called it "the eighth wonder of the world"
For its singular, majestically beauteous swirl.

< 10 >
— Instamatic Model —

Brains have parallel processors for form,
Texture, color, and depth—and a quick one
For motion detection—which all combine
Later as what we 'see' in unity.

AT THE SHORE

Where I am now, the sea
Is neither blue nor green,
But a color in between.
The deep dark hole
Of cold is not here,
Just the warmth aglow.

The ego is neither
Gone nor overblown,
But in equipoise, the known.
The calming waves roll,
Amounting here their toll
From the other side of the world.

I'm on holiday, on vacation
From my retirement...
Where might I be?
I am beside her,
Astride the duality
Of the yin and the yang.

There is no talk here of One,
Nor that nothing can move,
For all is moving life about.

So you know maybe where?
There is brightness all about
These shifting sands of time,
A heart warm beside mine.

No talk of me nor thee
Behind the veil of naught,
Just eternity's parenthesis.

The birds came down
From the sky
To pick the table dry
As the ghosts of Pacific
Walk the waves,
The captains of old,

For so it said,
As we read,
While laying in bed.

The wind on through
The curtains flew,
As I wrote some poems anew.

SICK OF LOVE

Whereof, and herein above,
I give up on endeavoring
To form rhymes with 'love',
And relinquish, for now,
Any mention of the turtledove,
Ring dove, or foxglove;

For love's rhymes are just too few
And it's getting just
Too darn hard to work '-ove'—
into that small, worn out space
Between God and Heaven above.

So, I'm through rhyming with 'love',
Having, anyway, nothing left to prove,
Or for that matter,
To more poetically improve
That sorry number of words
That sound like "love".

So, onward—to much better sounds I move;
Getting out of the rut and into the groove,
To hope, and wait for my poems to improve—
And, for my verse to behoove
That all love-rhymes do remove.

And, of course, by now, dear reader, YOU'VE
Noticed, and may only halfheartedly approve,
That, in England love' sounds like 'luve'.

THIS IS TRUE

This is true, my love,
That the lightning now flashes
In the light of your eyes.

The clouds in my heart explode
With the soft new beating of your own.

This is true, my love.
Your sweet lips become red
As a blushing bride's.

This is true, my love,
That the tree of life
Once again flowers within you.

Your mind awakens to follow
The guiding rhythms of my pure soul.
The night shall weep dew at
Your sight upon the Earth anew,
And the morning will surround you
With the light of life's delight.

This is true, my love,
That the touch of your warm breath
Intoxicates the being of life
Into and out of my spirit as well.

This is true, my love,
That the world will yet hear
Your sweet song.

Go forth, my dear twice-born one
To walk the earth once more
And help to bring evil to its knees.

Dame Fortune

Wherefore is the vaporous lady
Who rains good fortune upon me?
Her portal to the net must yet be broken.

Probability has regained its memory and spoken.

And yet a spirit in my feet
Still leads me -- who knows how? --
To her chamber window, sweet.

(adapted from Shelley)

TEN DAYS TO WHITEHORSE

A thousand miles through the northland white,
With its four peaks and endless trails of fright
I'll pass, with my dogs, four hours off and on,
Over frozen rivers and the tundra far beyond.

The moon will light the night like the day
As the winds whip my thermometer away—
Minus forty and lowered by the wind chill,
As I descend the mountains and the hills.

The Yukon is a quest to know what we are
And once were when we rushed for gold afar,
Flying past the furthest bar, over shone by stars.
What strength will I find where the eternal are?

What icy river might sweep me into cold embrace
Before I can finish this animal and human race?
What exhilaration may drain to solitary disgrace
In the middle of some unchecked pointless place?

What the whether of the extremes I can weather?
Whither whence I slide heretofore wench-untethered?
What fate awaits in this land of legend and lore?
Where no woman sledder has ever gone before?

I'm away and off from Fairbanks, forevermore.

AUSTIN DOES SPIRITUAL WITH RUMI

What flaming forge fires all that we know?
What do we seek? (I soft wondered aloud.)
We long for the TOE—as the human soul
Turns inward and out to find its way home.

Why do we wander around in the dark,
In the middle of the night like this?
Well, if I knew the answer to that one,
I would have been home many hours ago.

Where would that be? (I heard my voice then say.)
I don't know—mind ever seeks. Whatever
Which brought me here will have to take me home.
(Perhaps this is home and we're already there.)

How do we see this home from this new house?
Close both eyes, to see with the other eye.
Then how do we hear of it with our ears?"
The blossoms drop their blessings all around.

What quenches our thirst in this life of ours?
Break the wineglass, this earthly cup of thine,
And fall toward the glassblower's breath and drink.

Why?
We are the sweet cold water as well the jar
That pours it. Plus more—we are even
That which makes the drink taste so refreshing.
Where is the Light that shines to make me so?

There is a light seed grain deep inside you.
You fill it up with yourself, or it dies.
Where do we go to know, climbing mountains,
The Himalayas, to find the wise old man?

A mountain is but a little piece
Of straw blown off into the emptiness.
And what of her, the beloved beyond?
There is a window open in between.

How's that? The quiet airs mix our beings.
For, out beyond the ideas of wrongdoing
And rightdoing, there's a unified field.
Go forth and then wait; you will meet her there.

And then do we see the bright light of day?
This day that we seek is well outside of
Living and dying, sunrise, sunset and noon.
Do we not tire, always walking, looking?

At first, we did, yes, but then came a grand
Moment of feeling that wings had grown, lifting.
We fly? *The rhythm lifts us—music plays through.*
From. . .? 'Twas fashioned even before it was.

What do we feast on? *We taste the sweet taste*
Of eternity this minute. We're afraid?
We have long since wet our human robes
In the shallow water; we dive deeper.

We dive under, even naked under,
And deeper under the fathomless surf,
Wherein the drop becomes the Ocean, too,
As the Ocean, as well, becomes the drop.

(Later) Where have you been? (I asked of me.)
(I am thee and now you are also me)

Well, everywhere, and nowhere—in between.
I did not cease, though, from this exploration;
I experienced the world inside and out;

And the end of all my exploring was
To return to the place I started from,
But now I know the place for the first time.

TO CYNTHIA

Your figure is like a tree,
Bending with surges of wind,
Calling your arms unto me.

Your passions are unsinned,
The perfection of my fancy;
Of my ideals you are twinned.

Your spirit is of eternity:
All-pervading, never-ending,
Comforting in its certainty.

Your love rises on the wing,
Singing Heaven's rhythm there;
Vibrations sweep my heartstrings.

You're touching me everywhere
In all ways; within and without;
You fan my flames—they flare!

Your soft lips' sensual pout
Draws me into the depths—
Sweetwater puts my fire out.

Flames rekindle by your breath
When your breasts rise and fall,
As ripe fruit on the tree are blest.

Your eyes gaze—to me they call;
They are deep, glowing, bright;
Therein, I see forever and all.

Like the day snuggling the night,
Your being merges with mine,
Mingling in magical twilight.

Your visions of love match mine;
Often I have dreamt you up—
Now you're here, lovely and fine.

You're the elixir that fills my cup,
Love's essence distilled into being;
The scented breeze lifts me up.

This perfume is love fleeing
From you as you give it away
With kind grace all foreseeing.

Now take us to where we'll stay,
To the forest home built for us—
Where nature is and lovers play.

OUR CABIN

Come! I've built a forest cabin for us,
Away from the bustling cities and towns,
Where life's best things are simple and free;
See! We have air, earth, tree, bird, and sky,

A well, a porch, a fireplace, and a stove,
And water, diverted from a fresh stream,
Clearly flowing into our spring-room
And out, where it laves our garden. Look!

We grow entwined like honeysuckle twins,
Close yet free—two spirits as one become,
Living and loving among the thickets
That shield and cool us in Heaven's shade.

Each day pours life into our roots of love,
As flowers bloom and trade nectar kisses.
For such union a firefly shines its light,
And sparks the flames of a romantic night.

CELEBRATION OF SOULS

After love was made,
We, connected, stayed,
And, in each other's embrace
We laid, still in place,
While our senses melted away,
And were felt no more that day,

Having been replaced by a new sense,
A joy that lay beyond sense—
A realm of calm deeply felt
As everywhere it dwelt,
A sensation both mystical
And totally magical.

In it we drifted, crossing oceans
Filled with good emotions,
And floated down through
Deep caverns—deep we flew,
Rising and falling through a space
Where no thoughts could race,
Weightless, unlimited, unmeasured,
In the poetic land of many pleasures—

There becoming invisible, losing
Our bodily presence, choosing
To remain as one, although to
Even move would have required too
Much effort—of which we had none,
For, in spirit we had one become:

Ghostly phantoms, specters with
Human powers known only in myth,
Lying, awash, on some distant shore,
Our senses shining forevermore,
Like the sun, a scarlet flame above—
Beings quenched in the sea of love.

The pulse of love was still much with us
As we lay awash on the shore, resting,
Entwined, in the paradise of lovemaking,

Where, we rode upon the waves, receding
And returning, wet with liquid peace, fulfilled,
As now and yet again small wavelets
From the soul's ocean of emotion
Swept on through us, in ripples,
Echoes of the storm's mighty swell,

Vibrating and rinsing.
Waves seemed to come from within us,
Yet, from all around, relaxing us,
As each other we kissed,
While rivulets ran back into the sea,
Every drop tingling as it found us in caress;

Then another, and yet another drop
Quivered its waving way over us,
Cascading, while we yet embraced,
Connected all the while in one ALL,

Flowing, immersed in romantic afterglow,
Water sinking into the sands,
Half drying before wetting again—
Moisture rising up into the air
In one fluid motion toward the sun;

Then, yet one last whisper of watery sensation...
Calling us back into the sea.

< 11 >
— Screening Time —

Consciousness is referred back in time a bit,
Like the tape-delay of a live TV show,
To hide the brain's processing time from us—
Making things seem to happen instantly.

RECOLLECTIONS OF WAR

A fading eagle flew frozen in fear
Past deserted flowers in desperate land,
And a rising earth halted for a hasty madness
As time awaited a dead sun.

Remember now the beginning, one fine day,
When we came out of nowhere!
In no cradle birth, one thunderous heartbeat
Separated animal from plant,

And, we stood up straight one day,
Our minds still a drunk's uneven crawling;
Later, in the breath of life, we knew that
A churchyard must yawn now and then.

But now we are helpless—
We must fight to our worthless deaths, dying,
Screaming forgiveness, but, die as we must
When peace is a barren land.

Daily now, one grips less firmly his last integrity,
The essential life slips.

Where are the grown men, stuffed and rigid?
Where are they? Where?
They are so silent and meaningless to us now.
They are no longer with us.

And throughout the aftermath
We could almost grasp it in our dreams,
And hope that we might live to die
Far from the River of Perfumes.
Meanwhile, we are dying to live everyday
As we surrender our souls.
Around us we see the bodies—

They lie upon us; they died among us.
Rising to our last stand we look:
Where are the grown men, the old men?

We thought that we were loved then,
But we've been betrayed, sold, lost ...

Shall I try for fading woods,
Scrambling over the trails, searching for my life?
I'll flee and fly over the leaves of yesterday—
They crumble before my eyes.

And there I'll come out of it all
With firm desire to laugh, love, and live—
There in a hilly grove
Near swelling stream by daisies, grass, and tree.

Once more I escape the horrid death
As the grown men approach;
I try to see my way past
The swiftly moving figures of the human race.

Even now, those men with guns so loud
Are silently dying in the strife.
Living in a time nigh for sighing,
We rise for dying.
Can this be life?

Of course, all this it was our duty to bear;
We bled our blood; we served.
And during the lull of the monsoon rains,
I began to drink, to honor my life—
To hope, as dawn comes,
Much like a Chinese painting—
Too real to be true.

I wake the artillery-man,
And cross the Song Ba to disarm the claymores.

Now it is lovely April and we're dying
On this fine day in the time of our life.
Slightly sighing for crying Charlie,
My bayonet blazes in scarlet, in death,
And yet another hasty man
Gropes for the earth and escapes this horrid life.

But there, on a cloud of thought,
We fly by their ways with a life for ourselves.
And, then, they wither with the wind,
Those thoughts that once echoed,
Where they once were teeming, fighting...
The forums now emptied.

There is only room to say
"Let us kill him," as wrath's way becomes us,
And there in the cells of a brain
Where currents of feeling once surged,
The mind's will falters, and waivers
Between the Emotion and the Intellect.

A shrill siren chilled with ill will,
Then, when he was yet young and fine—
Houses were crumbling, streets were heaving,
People were weeping, dying;

And others wished to live,
From brothers to mothers—all lived but the father;
Can you see the tears in the young one's eyes
As the deathman cometh?

The love and the feeling were nowhere,
The men motionless and rigid,
And, too, the air was not worth breathing,
But, was filled and smothering,
Leaving the men breathless, helpless,
And, of course, so lifeless.
The blight was so deathtaking;
The sight of goodness never so breathtaking.

Once in awhile I'll wince in a smile for truth,
Cringe at the fringe of love;
It is my dream,
A star shining somewhere in the universe—
I can see it there in all of its dimness,
Through the plight of my brightness.

It is there forever and still;
It is there while the thinkers thought for ages,

As dreamers dreamt time after time,
When hoped even the hopeless,
As slept the sleepers into oblivion,
While philosophers pondered infinitum,
As wept the weepers for a long time,
When pitied even the pitiful . . .

All that I saw on Earth was lost.
There hated the loveless in the wasteland.
There the dying lived for a lifetime
As all the wise men greyed and died.
So now I'll let my 'enemies' grow old
As my wine yet flows sweet and pure.

Here comes the slush of doom seeping over us,
Belching with contagion.
The pleas of the corrupt fly out;
They cry out; their lives are snuffed out!
The Good Friday mourners yet weep for man,
For everyone, for eternity.

At life's end
The silent men array themselves, finally—
There for the asking
In the stead of the dead,
Prisoners of themselves.

Cautious Pilate ponders,
As there my star shines in the springtime of life.
The star is a beacon in the night of terror,
Fading in the search for the valiant.

How can I live, how can I die?
Look around—there are other worlds!
See, the grass is high and green
On the far side of never.

Find for me the sun shining, the streams flowing,
The forests, the fertile meadows.

The soldiers moved slowly now
To make their lives last,

A searching band;
And fighting has flared on the border;
Now hurry death or hurry darkness.

Deciding at last, I made an easy day of it,
Staring life in the face, indulging
In a vast wonderland and wilderness
Of childish fancy and fantasy,
And I laughed a lot louder then,
Feeling no need to weep in pity for them,
Or to cry for the scoundrels
Who would grasp at life from graves in war.

It was then that I saw the life,
The awe, the infinite,
The good, and my end.

To see where my youth and laughter could go,
I lived and died to be free;
My mind took no mind;
Yes it was good to be loved then,
To be young again.

< 12 >
— Time Framed —

The 'now' is ten-forty-fourths of a second long,
The frequency at which events appear over
The horizon of consciousness, the succession
Of which gives us the illusion of time passing.

< 13 >
— Out of Control? —

Do you control your thoughts or do your thoughts
Control you? Could you, silly as might seem,
Just be falling, hook and line, for your thoughts?
Think deep—thoughts may tell you the answer!

INFLATION THEORY

There was no place special in time
Nor properties of reason and rhyme.

Our beloved quantum fluctuations
Left their imprint all over creation—
The signature of their emanations
Written in the CMBR's variations—
A tiny magnifying glass upon their revelations,
As well as in the capitals of matter congregations
Of galaxies, nebulae, and other condensations.

What underwrote this glorious expansion
From such a humble state to a big time mansion?
It's called inflation.

Perhaps there are many such bubbles blown—
All but one of these pocket universes unknown.

Where did all this energy come from
To amount to this astronomical sum?
It comes from the gravitational field.

Our universe did not begin with this yield
Already stored in the gravitational field;
But, rather, the gravitational field can supply
The energy because its energy found
Can become negative without bound.

As more and more positive energy materializes
The forms of ever growing region sizes,
Filled with a high energy scalar field, arise,
As more and more negative energy materializes
In the form of expanding regions wielded
That are filled with a gravitational field.

There is nothing known that can place a border
On the amount of inflation that can occur
While the total energy remains exactly zero...

Why does this "zero" ever become the hero?

OLD AUTUMN

The glowworms, fairy stars come down to ground,
Gleam the shadowy woods through summer's round;
Then, fall's leaves flutter through the quiet air,
The autumn being the sunset of the year.

The rustling of the trees comes to my ear,
In this, the most mellow time of year.
The harvest brings fulfillment, yearning, too,
For autumn is both a smile and a tear.

Each year, in October, Jack-in-the-Green
Has a chilled rendezvous with Old Autumn,
Who colors the leaves that Jack made verdant
A season ago. They meet out in the woods—

Although never in the same place, for seasons
Come and go and meet each other as they may.
This year Old Autumn was a little late,
So Jack-in-the-Green sat down on a stump.

Jack pondered his disappearing green youth,
For someday he would have to take Autumn's place
And perform all of his withering tasks...
A few days later Old Autumn came by—

He gave unto Jack a cheery greeting
And a warm embrace that marked summer's end.
He gazed fondly at Jack, his younger self,
And saw the vitality that was once his—

Then said, "Once I was young; once I was you!"
"I know," said Jack, "Do you remember how
I refused to believe you, saying 'no'?"
"Yes," remembered Old Autumn, "very well—

Like the time I met the Old Man, Winter,
On a snowy December day long ago.
He told me that he was my older self—
But I didn't believe him! Told him off!

True, I was already feeling my age,
But, after seeing the old white-haired geezer,
I felt young again! Yes, he knew me well."
"Yes," said Jack, "so I made a little poem:

When younger, I knew not my elder same,
But when older, I told my younger same
That youth must be young—he knew not my name!
It was my younger self that was to blame!"

Swallows twittered in the skies as sprightly
Jack-in-the-Green picked a ripening gourd
And gave it to Old Autumn, who encouraged,
"You won't have to meet the Old Man until

You take my place, for only I can see him
After I take down the last of the oak leaves.
For now, the Old Man sends but his errand boy,
Jack Frost, your twin brother. Hi ho, here he comes!

Aye, young Jack, this is the rarest of days,
For the three of us can be together
But once a year on this bright day / cool night."
"The Old Man is so lonely, is he not?"

Asked Jack-in-the-Green, "for he sees only you."
"Yes. Old Man Winter lives cold and alone—
He never sees the fair maidens of spring
Who reinvent the natural world each year."

There is a chill in the air as Jack Frost arrives
And sings out a greeting: "Hello my brother!
Hello Old Autumn! It's going to be cold—
Our first frost, but don't worry too much—

It won't harm the pumpkins any at all."
Old Autumn sighed and quick replied: "Good.
Now the rest of the leaves will crack and fall
All the more due to the ice in their veins;

Yes, they'll fall like the illusions of youth,
'Lying carelessly on the granary floor' and

'On a half-reaped furrow sound asleep,
Drowsed with the fume of poppies', as Keats wrote."

Composing himself, Old Autumn continued:
"And for those of you who think that 'warm days
Will never cease', let us ever remember
Dear Johnny Keats who died so young (25);

However, he lived and saw much than some
Of us might hope to do in a lifetime."
A shiver ran through Jack-in-the-Green,
Hence he said: "It's cold; I must go now, for,

Summer passed away in his sleep last night;
Autumn, sweet and plump, carries his offspring.
The year dies in the night; ghostly winter looms;
Lo; the flower is already in the seed."

"Well done, young Jack-in-the-Green; quick, go, for
Soon enough comes your autumn of care
Sobering into age, thence into
The pale white winter of death,

Though not yet your warm indolent summer
Of contentment lazing into middle-age,
But surely past is our crisp,
Flowering youth-spring of joy!

Such then, comes the end of summer's dreams,
The blanching of the grassy banks of streams,
But all fragrances my elves remember
Through their long sleep in the winter embers.

The blossoms fall, showers of fragrant beauty,
As leaves fade, while the bulbs store up energy;
Nature's floral dreams grant this destiny,
For these leavings enrich earth's potpourri.

Flowers lay their heads to sleep in soft beds,
Blanketed by webs of gossamer threads;
My elfin creatures cast their spectral glow,
As winter stars—floral twins—start to grow.

Later, when surely all the world is dead,
An elf will stand atop Old Winter's grave
And say 'tis not dead, and, by magic bred,
Makes Snowdrops flower in the tomb's heat wave."

Once, I, the author, ventured outside at
Four on a dark frosty October morning...
It was so quiet that I could sense the
Cosmos as it played rhythm to my beating heart.

I saw a preview of the winter's stars:
Orion, you are so high in the sky—
There for only the astronomer's eye—
As all the meteors go flying by.

Then I heard a rustling sound in the leaves
Around me—a skunk perhaps—but no,
It was the sound of many falling leaves.
I knew that it must be him, Old Autumn;

He was out there somewhere. Then I sensed him
Going by, for some of the leaves on the
Tree right in front of me broke loose and
Floated away, hitting some other leaves

On the way down, making that rustling sound
That I'd heard earlier. Then it stopped, but
Soon it started up on the next tree, and
Then the next—and so I could very well

Follow the path of Old Autumn making
His rounds in the misty October morn.
Chrysanthemums drank the mellow day,
Falling petals carried the light away.

The weed-flowers grew, marking autumn's track,
The blossoms that almost brought the spring back,
But, winter's white death wrap was drawn over,
Smothering the earth's last warm sweet odour.

The autumn fog enswirled, the mist upcurled,
Into nothingness the wisp slow unfurled.

November flew by, a colorless dearth,
And December, amid death, a festive birth.

Youth and Beauty made agèd Winter mourn
For Summer's grain—the waving wheat and corn;
For Old Autumn, withered, wan, had passed on,
Leaving the earth a widow, weather worn.

Long since have the winds scattered the leaves
Of the trees to make of them a
Burial shroud for the flowers that died
Grieving at summer's passing. All is death.

The fall is now nearly lost to memory—
Winter is summer's ungrateful heir,
Squandering his riches and abusing his gifts'
It's not Old Man Winter's fault, but his duty.

Summer lies underground now, forgotten,
Silent and crusty, covered by winter's
Stern mantle. Only April's tears can make
His grave green again in the springtide.

As seasons pass, the world comes to our door:
Spring sings through the wingèd troubadour;
Summer calls with the rose, 'midst the woodlore;
Autumn crows, plump and sweet, through frosty hoar.

Joy and exuberance are spring's largesse.
Sunlight, warmth, and growth are summer's bequest.
Autumn brings wealth with the mellow harvest.
Winter's fruit is peace—its bounty is rest.

Past us is the flower of spring's soft breath,
Though not ended our summer of promise;
Soon enough will come the autumn of care,
Beheld, at last, the pale white winter of death.

March, April! spring!—we'll reign as we May there
Between June and her sister September,
Then prolong the fall, till November come
December, when we can sweet Remember.

In the whisperings of the after-years
The winds of time slowly dry the tears;
Nor would I take back a single drop, for
From those tears the flowers grew without fears.

In spring, we rise from the garden at birth.
Summer blooms long with the roses' fresh mirth.
Autumn creeps in—we wither on the vine.
Last comes winter, when we return to earth.

THE LOVE LIFE OF THE GLOW-WORM

Flashing desire, the glowfly twinkled across
The starry summer sky, love's energy unspent,
Searching through the darkness, with passion's might,
For the beacon of her consent—the mating call
Of pulsing, green and yellow light.

At last, came the reply:
"Yes, oh yes", a-light, she said;
Now he became a firefly,
As at once she did too.

To a closing flower they together therein flew,
Blinking, winking in the seclusion of its petal bed.
This dance of light and love—their honeymoon—
Brightened the night till it looked much like noon.

Those jolts and bolts, surging, merged in currents,
And swept back and forth as they signaled delight—
Fires luming and oft reluming the flames of love
With electric hugs,

For they had, by now,
Become lightning bugs.

NOSTALGIC NOTIONS

I ran my hand along a picket fence,
Counting heartbeats and running like a child,
Still carefully not stepping on the cracks,
Noting the furrowed ants bustling, thriving,

Wondering at a old chestnut tree that
Had somehow survived the blight, towering and
Ever so gently tilting the walking plane—
Presenting me with more ancient notions:

Of tire swings, swaying, hung from low branches,
Of a lemonade stand secure in the shade.
My youth came flooding back to me, into me,
And so I continued to give it life:

The back door of a bread wagon opened,
Releasing the fresh-baked aroma;
Mother came out with a handful of dimes,
Buying what would've taken three hours to bake.

On the houses' steps rested newspapers
And the sturdy rounded bottles of milk,
Compliments of Elsie the cow, truly
A vision from the grazings of childhood.

We played games on these walkways, like hopscotch,
Roller skating, and marbles. My bag of jewels:
A cool green cat's-eye, a big blue boulder,
And varicolored pockmarked throwaways.

There—a lush garden lovingly attended
By an old lady, accompanied by bees
And butterflies, all of which caused further
Indulgences in my flights of fancy:

As children—and now if we're young at heart,
We'd pause in play when that first butterfly
Fluttered by, that fragile ephemeral
Vision of something almost heavenly—

A flower floating on the air, perhaps,
Signaling that our endless summer had
Begun, that something called 'school' was now an
Artifact of ancient history.

Did the butterfly first arise from the
Soul of the pansy, before human times—
One of those edenesque transformations
That was inexplicably magical?

The metamorphosis is still charming,
Albeit but from a caterpillar;
Amazingly, delicate as they seem,
They flutter all the way to Mexico,

And take their sweet time, alighting here-there,
Meandering from plant to bush to flower:
We learn that there's more fun along the way—
The journey as rewarding as "getting there".

During our carefree days we'd swim the pool,
Diving off the side after pennies thrown,
Retrieving them from the bottomless deep
Near the big drains—then rising up, breathless.

Still at the garden, my mind back from flight,
The gardener beckoned me inward, and,
I leaned over the fence to smell a flower,
And a thousand memories reoccurred:

Each Morning Glory blossom lives but for
A single day, and is replaced by
Another, each in succession shining
In its morning glory, wilting in noon's heat—

Withering quickly in the afternoon,
Then languishing throughout the evening—
Their happy message to us being that
Another day will always come on.

The Amaranth intrigues—its leaves never fade,
Even long after death, ever remaining
Vivid red—could it be, somehow, that a
Portion of the infinite yet lives on?

There, the blinding luminosity of
Sunflowers; we dried the seeds and ate them,
Each still a glowing ember of memory
Of the bright days among a thousand suns.

I drank up Buttercup portions of the
Bright yellow light from the elfin goblets—
And entered the realm of fairies, pixies,
Fays, trolls, goblins, brownies, gremlins, and sprites.

We had cherries, and a grape arbor, too—
Eating them fresh, competing with squirrels
And birds, always forgetting to wash them,
Sour as they were, then spit out the seeds.

I walked on and saw a lake surrounded by
Old and broken down vacation cabins.
Of course we were never "there yet" when we
Asked, but soon dozed off, tired of asking.

We dug the worms at night, keeping them moist,
And got up with the sun to fish, and then
Skinned them, and cooked them for lunch or dinner—
This to me is America Remembered.

Dad was always out fishing—my brother, too,
And me less often. Now I clearly see
That fishing has little to do with fish,
But with cool breezes, moist air, peace, and quiet.

I wore my life preserver all day long—
Once I leaned over for a closer look
And fell in, swimming with the fishes,
Then pushed up, my life jacket now broken in.

We puttered to a mysterious island;
There we found—nothing, but camped and had lunch,
Feeling like pirates, and telling no one
About it until a whole day later.

At night we watched the bears forage for scraps
At the garbage dump; however, one night
The bins were empty when the bears came out—
Then they all turned and looked over at us!

Mom used to say "Come in out of the rain",
But nowadays, the sun is dangerous,
Unless we wear sun block, so she says,
"Have enough sense to get out of the sun!"

After a storm, when the sun returned,
We'd run out to see if there was a rainbow—
That shimmering otherworldly vision of
The colorful secret of simple white light.

How are colors made from three primaries?
Why is the sky blue? What unknown colors hide?
Well, color was invented in the 60's—
Just look at TV shows made before then!

To keep cool we once carried pinwheels, fans,
Parasols, and sucked on a piece of ice.
Now, with TV, internet, and cool air,
We stay inside of the house all day long.

By eavesdropping on the party line, we could
Hear real scandals and idle rumors, and,
If it was more interesting than watching
The grass grow, we stayed to hear the whole story.

Before the invention of the telephone,
All was conveyed by tell-a-woman, but, now,
We only answer to computers, saying,
"To talk to a human being, please hang up".

The corner market carried everything—
Eden's shiny red apples calling out,
"Touch me, take me, eat me", and, soon, trouble
Was at hand but it was crispy, sweet.

I rode my bike everywhere—I always crashed
On the killer hill, on roller skates, too;
Now I drive my car there, carefully—
Yes, I'm finally getting over the hill!

Always picked up a penny for good luck,
And pins, too, for even more good fortune.
I found a horseshoe all of the sudden—
'Twas bad luck 'twas still on the horse's foot!

Rural cemeteries were parks, too, back then,
So we played near the duck pond, giving them bread.
Some years later I returned, like a duck
That had been away for too many summers.

There were monkey bars for the climbers, and
Seesaws tottering for the restless, and
A refreshing sprinkler to cool off in, but
There was always some kid sitting on it.

We made greeting cards, keepsakes, with ribbons,
Lace, assorted scraps, and original words.
Now, we buy ready made cards with fluffy words—
In a day or two they are in the trash.

Simple pleasures are as free as ever:
The sights, sounds, and scents of nature; picnics,
Reading, writing, giving, riding, playing—
It's hard to ever get bored, isn't it!

THE RESTLESS WIND

Rising slowly from the cold dark hollows
Where the night airs fell and soundly slept,
The restless wind left her secret bower,
And, gaining strength, lovingly surrounded

And caressed the willow trees, which wavered
And swooned in her wake, as she, the wild and
Wandering wind, flew by in a cool breeze
From the west on her undulating wings.

Spreading the incense of the morning to
Nature's world of growing and living things,
She woke the flowers from their slumber
By drinking from them their blanket of dew,

Then told the tales of the joyous forest
To the birds, who soon carried them aloft,
Thence into my ears: songs of streams flowing
Freely, and stories of a glowing sky.

That promised many sunny hours to come
In the dreams of those who felt her passing,
As sleep was washed from their languid eyes
When they sensed that new dawn arriving...

As if some transparent veil had lifted—
When she gently stirred the embers of the
Last watch-fire and whispered softly to them
That the stars had gone and day had begun.

< 14 >

— Thinking About Thoughts —

We fall for our thoughts, hook, line, and sinker:
Conditioned responses, reflexes, or
Overwhelming emotions, spurious,
Or ancient, planted by evolution, or unbalanced.

ON EARTH

The sun fills the waking and breathing world
With the fire of her imagination.
In poetry, the sun is the power behind the mind;
The moon, planets, and stars are symbols, too.

Sometimes intellectual beauty is bright
And ideas gush from the eternal flame;
Sometimes it fails when the shadows of clouds
Dim the clarity of thought now and then.

Quenchless, boundless, ever bright and burning,
The mind's light searches every dark cavern,
Probing, imagining—its beam alighting
Upon the earth or high atop cloud mist,

And melts, with heat, energy, and desire,
The fog of lone reason and pure passion,
Burning it away, soft dissolving it
With the love of life, earth, mankind, and star—

From which comes adventure, friendship, delight,
Joy, success, triumph, and lasting gladness
Throughout the sun's journey into the night,
When stars shine on mind—suns they also are!

The moon fills the sleeping and breathing world
With the icy coolness of chaste reason
Unaffected by deep burning passions,
Although sunlit to glow in its wan light.

Reason, unsteady as the variant moon,
Oft does not rise in the night to guide us,
And deserts us in darkest times of woe;
We are alone on a black cloud-bound night!

Else the moon hides in the bright light of day,
Or is lost behind an overcast sky;
But, moonless nights take us beyond reason
When the stars excite us with their lights.

Yes, inspiration returns with the stars—
A thousand ideas beckon from afar;
Ideas wink like fireflies on the mind's meadow—
As starlight they stab the darkness of nought,

Until star-like Venus rises near dawn.
Goddess of romantic love and passion,
She captures us within emotion's swell,
While comets flash and confuse the wild sky.

Soon intellectual beauty returns,
Borne on birds' wings as song into the dawn,
For, all human music is but a part
Of earth's ancient melody and rhythm.

Imagination now soars past a day,
And into the season of spring's fast growth;
The shade is deep and cool, like the ghost of
Winter passing—gone but still remembered.

SEASONINGS

Nature Springs from Winter's tomb,
The bloom already in the seed,
The tree contained within the acorn.

Surging sprigs sprout from the soil;
Spring showers make the Summer flower.

Summer wakes from Spring's dying kiss,
Blooming when the rose does,
Sunning after the Spring's running.

Summer reigns upon the land,
Eventually fading in the night.

Autumn Falls as Summer leaves,
Harvesting its sum of days,
Seconding the rose of Spring.

The smile meets the tear—
Fall's embers last through December.

Ice winds stalk the weed flowers,
The ghosts frosting the dead stalks,
Snow crystals barring all that grows.

Winter is death cooled over;
Melting snows feed Spring waters.

< 15 >
— No Real Choice —

The brain's decisions are determined by
Memories, associations, and
Learned behaviors right up to the instant—
So, our 'decisions' are predetermined.

< 16 >
— Undetermined Will —

The 'free' in free will has no real meaning,
Unless we take it to mean 'random', that
The will depends on nothing but dice rolls;
What good would be such a brain anyway?

< 17 >
— The Effect of Influences —

Can you start or stop your thoughts? In other words,
Can you will that which does the willing? Try it.
Oops, a surprise thought just came from the blue;
You did not will it—the will is unfree!

< 18 >
— Your 'Decision'? —

A hormoned hunger pang, midway between a
Pain and a withdrawal symptom, makes you
Run, run, run to eat—well, that's OK—until
Another hormone signals satiation.

< 19 >
— Hormoaning —

Bonding hormones bring us closer together,
Pheromoning into lust, love, and relationships
Spurred onward by love-made endorphins—and, so,
Yes, there must be chemistry in coupling.

< 20 >
— A Thousand Minds —

The mind is perhaps many little minds,
Each a simpleton awaiting control,
Such as when we eat, socialize, or fight,
None of them very complex at all.

< 21 >
— Survival of the Fittest —

Subconscious trains of thought vie for attention;
The dueling choirs compete for first place
In the mind's 'I'—consciousness—to produce
Future, for this may be the task of thought.

< 22 >
— Awareness Afterwards —

The brain, with its hundred billion nerve cells,
Does all of our decision-analysis,
Only making its results known, at the last,
To the mind's highest level: consciousness.

ELFIN LEGENDS

Of man and angel, one yet neither, they came
To dwell forever in shadow worlds between
Form and substance, they, all elfin creatures
And all who float or fly as came from Paradise.

Yet neither here nor there, though everywhere,
They're the fairy host, nurslings of eternity
And of all things everlasting, like Amaranth,
And of all things Heavenly, like love and dreams.

Alive only at life's Heavenly cusp,
They appear but in half-light dawn or dusk,
Seen usually by some quick sideways glance,
Or through some autumnal haze, perchance.

Fays live in a moon-blue star-mist, out of space
And out of time, rising each morn as vapors,
Wavering here-there, in rainbow colors,
Being the light and life of all leaves and flowers.

Midwives of bloom, wizards of natural miracles,
Painters of green, and guardians of buds,
They, the keepers of what moves all things,
Are, with flowers, Heaven's smile upon the earth.

Born of kisses, fays are life's spirit-soul,
So much felt as to be oft seen and heard.
Their musical wings play songs so intense
That they fall as perfume upon the sense.

Fairy tinklings are sensed as drowsy fumes—
Incense lifting one on wings of fairy sighs,
The tide that turns us, as seen in the wake
Of leaves rising in their swell on windless nights.

The gossamer mist snares me, barely felt,
A fairy cloth of prehistoric weave that yet
Haunts every stream, meadow, and wood
In the Land of Youth, where timeless beings live.

For, as I'd sensed the cloud of lilac fragrance
From a mountainous bush, that passing mist
Awakened ancient creature things in me
That sympathized on an old frequency.

To life's forgotten tides and swells I yielded,
And, thus, was allowed into their spaceless world,
Through a small opening that tunneled, at first,
Then funneled into the expanse of Fairylande.

The Lande appeared, at first, much like my own,
But, I saw colors that I'd never seen—
That were neither blue nor green nor in between
And, further, they shone in some strange direction.

So this tale I give you, if I can return to tell it,
Of shadow worlds within earth's dominion,
Preternatural places transcending time and space—
The enchanted Faery world, Earth's missing link.

Other mortals I saw, too, while passing through,
Then knew that Elflande overlapped our own;
They were not asleep, but frozen in motion,
Awake yet unmoving in their instant of time.

Such, when each moment passed unto the next,
These mortal beings passed, too, wondering what
Might have been seen: phantasms in the mind
That fell between the frames of their living film.

Trumpet flowers had announced my coming,
My ticket being the poems that I'd written
On the lore and legends of the flowers—of Eve
And elves bringing forth all that bloomed and grew.

All things I felt continuously now,
As in humankind I knew only in
Rare moments of ecstasy when melded
Happenings had lifted me heavenward.

Magical things I saw—that only appear
On earth when one's eyes close but for a second:
Wingèd ladies, and flowered butterflies
Whose prints are pressed as dust upon the pansies.

The birds were of a species never known
And seemed to share a special closeness
With their elven brethren, faery sisterhood—
Which I knew and felt and saw as kinship.

I heard woodlands that once only whispered,
Meadows where there was once but a murmur,
And grasslands, unhushed, full of wondrous sounds—
The music from beyond the human range.

My senses were heightened: touch went deeper;
My eyes saw colors beyond the spectrum;
I reached into living things, knowing them,
And the odours called, mixed with emotion.

A flush of youth shot through me, as the chain
Of light from angel to faerie added my link,
And my eyes were sparks of bright burning fire,
Sense extended in a new dimension.

I sprouted wings and flew, like a bumblebee,
And fell in love with a lovely wingèd flower
That had come to life, a vision of fantasy,
Her elfin eyes beckoning me toward ecstasy.

Summer follows us around, elfin queen
And I, as we lay snug in winter green,
Your glowing pixie crown lighting the scene,
Your curves spooning, the ears pointing away.

As fays, we made love in the air, hovering—
Evanescent visions of disembodied happiness,
The magic link in the chain of things, connecting
Man to God, by angel and star, to all that we are.

Satisfied, fulfilled, yet desiring more,
We returned to our cabin, loving deep
Into sleep, as blackness fell all around,
But for the starry memories that glowed.

Although I can stay no more than a year away
Or lose mortal form, this place will be my home,
So, here I'll return, the seasons going round,
Where I'll continue to expand this poem.

Although back to deliver these words, I already
Miss the sylvan solitude and the crystal pools
Of enchanted worlds between Heaven and Earth,
Where the wanderers of light call me home.

So, now, live your life and dream a dream, through
Dale, meadow, and field, in grove and greensward,
Across love's pure stream, in shimmering sheens
Of the dells of Elflande—it takes but a wish.

< 23 >

— Mindless Acts —

People act, robot-like, since they know not
The 'why' of what they do, for decisions
Are made 'blind', by brain networks, just before
They're presented to us in consciousness.

< 24 >

— Veto? Or Next Thought? —

Consciousness comes three hundred milliseconds
After the brain does its analysis,
And thus has but last second veto power,
If allowed, over what the brain comes up with.

THE VERSE

Although the day-tide had barely spoken,
He, nonetheless, opened their precious token—
A mysterious book of poetry that had been sealed
With a waxen shield, it remaining concealed
For over ten centuries in the secret chamber
Of the library of the old monastery's remainder.

The tome was written in some foreign language,
In verses of thirteen syllables in four-line stanzas.

They opened it as one would a tender lover:
A small bottle was encased inside the front cover;
Some of its spirit had apparently escaped
When the volume had been undraped,
For they'd been captivated by the Persia fumes—
The perfume of ageless rhymes from ancient looms.

"It's written in Persian," she noted, looked,
Having handled many of the foreign books
In her role as editor in the abbey's nooks.

"It's the library's most valuable book,"
He said, having illuminated and unhooked
So many of the monastery's great books.
"It was the only one I could save;
The only book we'll ever crave."

They watched, amazed, as the book came to life,
Like a good husband in the presence of his wife.

The words of the Persian poems then began
To move around the page, as over it they ran,
Sometimes briefly changing into English,
Entire verse-lines dancing like a dervish.

Then, after settling down from the struggle
The words would yet again jump and juggle,
Hanging back and then ever surging forth,
Darting around through the verses' course
Within each stanza to form a brighter source,

Lines which yet stated the differing aspects
Of the original and pervading concepts.

'Twas as this magical language transmogrification
Was attempting to preserve the entire relation
Of the original poetic scheme throughout—
The whole translation process so devout,
Including literal meaning, rhythm, rhyme,
Melody, syllables, meter, and time;
However, this didn't seem to be workative,
And so it followed that something had to give,
And that 'something' was the ration
That was usually lost in the translation.

Finally, out of apparent desperation uncaged
The Persian verses jumped right off of the page
And splashed into the bottle of perfume,
Wherein they redistilled themselves, subsumed,
Leaping back out and on to the empty page,
Whereupon they recondensed, restaged,
And recomposed themselves for this new age—
Into Victorian style verse—into new quatrains
In which only the essence of the remains
Of the original concept of meaning was maintained.

The lines were now ten syllables, rather than thirteen,
With so many related meanings heretofore unseen;
But the verses were still in groups of four per stanza,
And the correct lines still rhymed, yet per lingua,
Although some of the rhyming schemes
Didn't always have quite the same means.

Yes, some things unnecessary had been lost,
But something new had been added and tossed—
Something somehow much better told,
Although still within the spirit of the old.

"What are you?" she asked of the book.

"Are you alive?" he asked, as he shook.

The book replied, "I am the book of life,
My pages rife with the antidotes of strife;
I am conscious dream, a living philosophy—
I live forever through my words, wholly.

"On my pages you will find all of man's follies,
Joys, sorrows, wisdom, and all the jollies.
Read me and my ideas will come alive—
Demonstrating the happiest ways to survive!

"It is by experiencing my words
That you shall know them forwards.

"Yes, many arts may enrich human experience,
But they're no substitutes for the living of it."

< 25 >
— The Illusion of Control —

Decisions are not made by consciousness,
Although, this fine picture in the mind's 'I'—
Merely the brain's perception of itself—
May be fed back for further analysis.

< 26 >
— Unconscious Selection —

Not much of what the brain does reaches
Consciousness, and even when it does,
The mind's last to know—it's like a tourist,
For decisions precede their awareness.

< 27 >
— Machineful-ness —

Humans are like machines, going the way
Of their brains, genetics, and chemicals—
But, you, learning these secrets, rise above,
And at least know that you are a machine.

THE WAN MOON

Darkness drains my life away,
Sickness consumes my spirit;
My mantle is heavy lead,
Life's last glow is upon me;
My eyes are craters gone dim.

Death's ebon form seeks me out,
He covers me with his cloak.
"Come away with me," he says,
As he cools my burning brow;
"I offer you quiet peace."

A sudden strength comes to me,
In my waning crescent wisp.
In night's cold shadow I say,
"Un-hold my soul, Moon Reaper,
I shall fully shine once more!"

THE YEAR

Hail!
Winter storms the Year
In the month of Brand-new-airy,
Then Feb-buries us in snow!
March, Lady April! Spring!—
Let's reign as we May
With sum(mer)maids
Named June and Ju-lie,
Until, after A-gust of
Hot withering wind,
The sunny fire burns out—
'Cept embers, when
Leaves Fall into Oct-tomb-burr—
Till—no leaves, No sunlight,
No sky, no warmth—
No-vember!
Next de rain, de sleet,
De cold—De-cember,
When all that we can do
Is but sweet Remember.

THE TOEQUESTORS AND THEIR NATIVE LANDS

Melanie comes from a U.S. satellite,
A land of mist and not driving right,
Called England, she having been born there
As the Loving Goddess somewhere
On the sacred Isle of Women—
And then taking off swimmin'.

In her country, 'left' is whatever is right,
Leaving 'right' to be whatever is left. Right?

When I went over there,
I always had to use a mirror.

Melanie uses both brain hemispheres,
Mostly in the pursuit of achieving there
The brainless bliss of nothingness,
As of the Bra-Man's dreaminess.

England is not as bad as Graybeard's and Tina's
Upside-down provincial dominion extrema
Of the Mars-like planet of Australia
That has a bunch of dry dusty towns
Where they say "What's going down?" each day
Instead of "What's up? Hey!", but, either way,
The answer is always an elevator or a lift.
A foreign lady at a hotel once asked me for a lift,
So I picked her up and carried her, quite miffed,
All the way to her room, as a welcoming gift.

Down Under: I flew to Australia only once,
On assignment with Bill Bryson's penal sentence.

Australia is a mostly an empty thralldom,
Much more so than even an atom,
And is extremely far away from everywhere else.

It has less people than Tokyo,
But is a zillion times more extremo.

The constellations are the inverse
And the seasons run in reverse.
It has nothing of any interest
And the climate is the cruelest.

It is the only country that is also a continent
That is also the world's largest island extent
And the only one that began as a prison—
Graybeard must have a lot of original sin
Because all of his ancestors were his criminal kin.

The cities are all on the coast since the interior place
Is an endless desert about half the size of outer space.
Somebody once set off an atomic bomb
In the Great Victoria Desert's silent calm,
In Western Australia, a land much embalmed,
And no one even noticed it for years—that maelstrom,
But for some creatures jumping right out of the genome.

Of the world's ten most poisonous snakes,
All are Australian, for Christ's and God sakes!

Even a fluffy caterpillar can kill you;
Seashells are venomous, too;
Adieu and skidoo to you.

Every ocean current carries you far out to sea.
They even lost a prime minister, Harold Holt, see,
Who was merely strolling along the beach.
He stepped into the surf, going swiftly out of reach,
And was never seen or heard from again, "impeached".

Australia is very old and nothing has changed
There for 60 million years, nor anything rearranged;
Thus, one may find the oldest fossils on earth.
Even the first faint signs of life can be seen: it's birth,
And the earliest animal tracks ever made; no dearth.

It seems, though, that its creatures evolved
Outside of Darwin's theory, being all convolved;
They don't run at all, but just bounce
Across the landscape, like a ball, or jounce.

Australia is the driest, hottest,
Most useless, infertile, flattest,
And climatically unbalanced
Of all the continents instanced.

It is so hot there that recently, amen,
The air caught on fire once again.

The place is so inert that even the soil is a fossil;
Even the worms and bacteria are quite docile.

On the up side, they have 20%
Of the world's slot machines present
To serve less than 1%
Of the world's population extent.

As about equal to finding a live T-Rex,
Proto ants were even found alive, having sex,
Although nothing like it had existed on earth
For over a hundred million year's worth.

So, does Graybeard know his evolutionary stuff
Or what? Yes, indeed, plus all of those ants so tough
Were found on his back porch, hellbent with intent;
They have now become extinct due to his experiments.

I flew back to Los Angeles from Australia, getting there,
In time and date, even before I left there,
Which was hardly soon enough for me;
Let us all say a prayer for poor Graybeard to be.

Science News: NASA is feverishly pursuing
The manned Mars Landing plan, fliers ever wooing,
But still needs some suitable astronauts,
So off to White Cliffs Australia for some kumquats.

Here they found a population of about 80
In a wilted world of heat, rocks, and dust, matey.
Due to the horrible heat, their cave-houses
Are burrowed into the hills—for souses and mouses.

When a vehicle goes by, it raises a big cloud

Of red dust that eventually settles, enshrouds,
And covers everything in sight, leaving nothing to see.
They have had electricity only since 1993,
TV since 1998, but no channels yet to espy.
Taipan snakes slither by, on the sly,
Their venom 50 times more poisonous
Than a cobra's—to leave you breathless.

Australia began as a nation thrust
When convicts actually began wanting much
To go there for the crazy gold rush.

So, anyway, NASA enlisted them all
In the Mars Landing Program's shortfall,
Figuring that they wouldn't really know the diff
Between the planet Mars and White Cliffs,
Or, that, if they ever did, they'd be so spaced—
Happy to reside in a much more hospitable place.

Science Saves Us From the Warm Future:
The White Cliffs Underground Motel moochers
And those of the various home residences picked
Have the right idea to avoid the warming conflict:
Free cooling to 67 degrees F: perfect!

This may be a good plan for all of the future deal
If global warming really happens to happen for real.

I can imagine a stay in the Dug-Out Motel, unraveling,
It being quite a heavenly destination after traveling
Forever going over bumpy roads and then getting
Out of the 'blender' mode and into the pool, wetting,
As all this traveling would grant more appreciation
Of the three star AAA motel's accommodations.

Coming, sweetie?
(Just you and me.)

Our room would have natural light from a shaft,
Which saves on electricity, oil, and gas.
There are no windows, but that only saves us
From having to view the non-scenery—a plus.

Cell phones wouldn't work, but, hey, no endings,
Then, for those blendings never ending;
There would be no interruptions
Of any pending eruptions.

Hey, how come people in soap operas
Always answer their darn phone; sagas?
It always ruins the moments erotica.

Plus, we could always dine in the restaurant,
Since the nearest supermarket is a scant
And rough six-mile drive away, askant.

And those dust-assisted sunsets
Are of truly unbeatable descents,
They having ten times as more
Colors than the rainbow: fourscore.

Plus, with White Cliffs having electricity now,
The beer is no longer like a steamy hot cow,
At 110 degrees F, but ice-cold, for highbrows.

There was a bad drought in the 1890's here
And the land has not recovered, oh dear,
But who needs that when one has love & cold beer.

In Fredrick's pyramidal world, it is that the origin
Of the belly button of the universe was an 'inner' begin
That reached the limit of being really small
And so it popped back out became the 'outer' all,
As in the outer space of the entire universe around
That formed from the unlimited merry-go-round.

Mohan wrote many multi-verses of poems
On the steamy planet of India, per diem,
Where they have three million Gods become.
They will eventually be getting more
So that each person can have one to adore.

One day the temperature there went down
To 75 degrees F, which would be a perfect markdown
Anywhere else, but here they all tried

To look for sweaters to put on outside,
But they didn't own any
At least not very many.

In the summer, on and in the icy planetoid
Called the Yukon, a frozen place that's best to avoid.
LabelWench worked only at night,
But, like the day, it was just as bright.

The sun never rose, staying up all the more
For it had never set on the day before.

When a cloud came by, in its starkness
They called it night or darkness
And had to use flashlights until it passed,
So that they could all find their wineglass.

They have only two days a year there,
Each six months long and longer,
Called white and black
Or bright and pitch black.

MJA lives in a land where there was no
Difference in anything, for all was equal.

Everyone was a clone, wearing the same clothes.
The sports results were always ties—so close.
Everyone got an 'A' in school,
For they couldn't measure the rule.

But they'd all progressed beyond this equality,
Thanks to Bottomlander, who lives in a valley.

Antonio lives in the celestial body of Mexico,
But for some reason they call it Texas, since long ago,
The U.S. captured it. We would give it all back
But for the fact that they have already taken it, Jack.

I always try to say 'Remember the...', the start,
But I usually can't recall the 'Alamo' part.

One time I got a letter, for my laxes,
For Austin, on taxes
From Austin, Texas
And I didn't know what to do but axe it.

In Greenbug's asteroid of Greenland,
Every single thing was green, and,
So, after a while, this gave him the blues,
After which valley of depression's dews
He then felt very much in the pink,
And much on the uprise until all was a rosy think.

Then he discovered that he had been given
Green contact lenses at birth, these being riven;
His land was really all white,
The 'green' in the land name's write
Being only part of an advertising plan
To get people to settle the land
There after not so many came to Iceland.

Bogie resides in the sunken land of Florida
Where the year-round heat is all too horriba,
Where old people walk really slow in front of you
Towards God's waiting room, to the very last pew.

Bogie cools his thoughts in the arena of Tampa Bay,
Pondering every thought that comes his way.

RascalPuff lives in Niihau, Hawaii,
A secret place; so, that's all can say I.

Graham lives in the clouds where all is allowed,
In a levitated home, smoking pot homegrown.

Felix lives in Schrodinger's cat-house shed,
But only half the time, when he's not dead.

Lloyd lives in the real house of science,
So please let all posts there be in compliance.

Leskey's leaving the land of zeal,
Ever becoming more and more real.

Max lives in the U.S. in the state of Deep Thought
With all his relatives, many of whom were fought,
Cousins twice removed, but they kept on coming back.

Melanie says that "Nothing is Real"
And this reminds me of guy whose spiel
Related that "Nothing is true". For real!
Everyone believed him for 20 million years
But then they found out he was lying—oh, tears,
And so it was then so sure that they truly knew
That the case was really that nothing was true.

And so for 15 million more years of bluffing
They believed not anything, and nothing.

So, what was this guy's name?
Well, it was the man with no name,
Which was Nobody Nowhere
Who was now here.
He lived in the Noplace
Of virtual space.

Then a large but tiny problem
Was found with nothing, ahem,
That there was a slight ado about it, surely,
This being the quantum uncertainty,
Not a very big deal, really,
Being the smallest thing of reality,
But enough to raise it to be a near nothing,
Just about as close to nothing
As one could ever get, without stuffings,
And so it hardly really counted for much,
But it made for a universe in which, as a crutch,
The gravitational energy was negative,
It canceling out the positive—
All the energy of stuff,
But for the unavoidable touch
Of the quantum uncertainty
Which we can almost certainly
Avoid for all practices, purposely.

So, "Nothing is true" and "Nothing is real"

Turned out to be pretty much right, a done deal,
Except in England, where it was all that was left:
Reality bereft, a cleft from the theft that was deft.

Meanwhile, Mel shivered with the quantum jitters,
Turning it into a jazz dance of some random twitters.

Now, what about "Nothing is real",
Employing it in the sense that the real
Doesn't even exist, although it has a feel.

Well, something does exist,
So we might rather say that nothing persists
And, so, "Everything is temporary", but being,
Since our realism came from a near 'nothing'
And to such it must return, in great arrears,
Even if that takes about 10**10**10 years.

ProfPat is expanding into the void, accounting for
The Catholic girls' heavenly student bodies more,
And lives in the naturally divided state
Of Michigan, that unmarried state
Which is separated by a long fat lake,
A further segregation being that the upper part
Of the state associates with Canada,
While the lower part is called "Michiana".
On the other side of the lake the state
Is more or less a part of Wisconsin's fate.

Up Above: One time I drove back from Chicago's exploit
To New York by entering Canada from Detroit
Into Windsor, near the Church of ProfPat; adroit,
By then getting out of Canada's shortfalls
As soon as I could, at Viagra Falls,
A fun place for vacationing foreigners
To leave lots of money for souvenirs.

When they say that the glaciers retreated,
They only mean that they repleted
And went back up into Canada, where they sit
Atop the forgotten land, in which mitt
Few are cold because so many are frozen stiff.

Since it is all ice, everyone plays hockey all day,
At least when they can blowtorch the ice away
From their igloo-cars, if the flames will stay.

Canada has only one super highway:
It goes east and west all day
And as close to the US as it possibly can.
It doesn't even have railings, man,
For there is nothing to run into
If you go off the road by some miscue.

So, anyway, Canada is only really only about
A width of ten miles that is a barely habitable shout
Just above the U.S., a suburb really, for fallout.

They have only one baseball team
That is going nowhere, it would seem,
Since they are really all hockey players extreme.

The police still ride horses there, on the loose,
But this is actually a step up from a moose.

Sears is their biggest industry, barely afloat,
Mostly selling really heavy fur coats
Made from polar bear furs that were poached.

Mikal works in men's clothes there.
She drank Canada dry in her time that was spare,
But now drinks only ice water, right from her tap,
Being sober and serious and all that yap.

All their restaurants are called Tim's Donuts.
Canada is really even smaller than it looks, a rut,
For a part of France called Quebec is in it, but,
Is not really with it, they all being nuts.

Every US map I've ever seen ends at Canada,
Just showing it as a bit of a blank gray area,
Which probably is really the right sceneria.

When Henry Hudson discovered the frigid Hudson Bay,
His men were so mad that they put him off one day.

Mikal lives at the end of long Lake Ontario,
A prime spot since it gets the extra lake effect snow
Of two feet more than the average of five feet or so.
June is still a winter month there, in that fen,
But one only needs two coats to wear then.

One time it got up to 80 degrees F there
On a mid-summer's day and everyone there
Became sweating and boiling and so there
They all ran around with naked eyes, all bare.

Even the Yukon, which is really a secret part
Of Alaska, has better weather than Ontario's heart.
The power goes off more than it's on.
The only place worse is Antarctica wan.

So, to help Mikal, let us apply some science
In the form of some secret zapping rays of potence
From the North pole's array that will melt all the ice
And then flood Canada ten feet under; very nice.

"What's the big news from Canada, Benny?
They don't have any, Penny."

As for those in the rest of the world orientated
It is that the Easterns are dis-oriented
By the Westerns who resist the other orientation.

< 28 >
— The Variety of Robots —

We're all robots, but, no one notices
Since there are so many different kinds,
Which, though it makes life quite interesting,
Obscures the fact that the will is unfree.

THE SECRETS OF THE NIGHT

Soft and warm, the evening caresses me,
In gentle darkness and quiet stillness.
I beg her to yield her dearest secrets,
To reveal the full truth of what she is.

Much I already know from twilight dreams,
And from poems unveiling truth and beauty,
But, I ask, with my most inquiring looks
To know the deep mysteries of the night.

Above me, fires burn the stars away;
Below me, the Earth turns under my feet;
Within me, unworded dreams haunt my soul;
Around me, night pours blackness on the ground.

Often I've deeply felt thee, phantasm,
Known when you were there to encourage me,
Felt your touch in my heart between its beats,
Always sensed your presence in the mind's sight.

Now I ask from your powers of the night,
Not immortality, nor youth, nor birth,
But only that I retain your presence
Within me, in rhythm and resonance.

Now I sense your sweep across my heartstrings,
For I'm undistracted by day's bright noise.
NOW I hear your voice singing with my own;
NOW I know the love and goodness of man.

THE GARDEN

What once I was has dimmed, physically,
But, I am a star, still bright in the night,
Though, when the sun rises, I disappear into her.
For, no one looks for the stars when the sun is out.

THE NEW CHURCH

I was going through the channels a while back
To get to the Olympics and saw a Catholic priest
Saying and preaching
*"If you don't believe in what we believe
Then you separate yourself from the holy body
And blood and you know what that means.
Well, we really don't like to talk about
The consequences of heresy but..."*

Here's a much better idea:

For a good and better life, join the Weed Church:

The Church of the Weed

Weeds are the most energetic and alive things
On the planet and therefore we can channel
Much of this vigor into our own selves.

Costly vitamins and energy drinks
Are no longer needed,
Weeds being free and abundant.
Trust me. Have faith.
Weeds have the super power
Of nuclear energy
Without any of the side effects.

Some weeds have roots three feet long!
Even one molecule left of a pulled weed
Can regenerate the entire plant.
Weeds are of the unlimited power.

The weed is the Theory of Everything.
This is the greatest discovery of all time.

Our focus is on plain old weeds, not pot,
But no denomination of weed is excluded
And so the incense smoke spread that's about
Would most likely be pot,
But I can't say that here, if it is.

This stuff may slightly drain some energy at first,
But then the munchies will take over to allow us
"Munch" more ingestion of our special weed salad.
However, if anyone turns into a stoned statue
That's really OK since it's perfect for a church setting.

The social hour and the church hour
Are exactly the same thing,
But we will occasionally say things
In honor of weeds and will have
All varieties of weed pictures on the wall.

There is a fairly new 'mile-a-minute' weed
Going around New York and so this
Will be the centerpiece of the altar.
Even the pews and chairs will be
Made out of woven weeds—
As a constant reminder of the glory of weeds.

The floor will look like the average person's lawn—
Mostly weeds and clover and hardly any grass.

The weed salad will be washed down
With dandelion wine and will provide
Such great energy to us all that we will become
Much more alive and creative,
Perhaps even posting 20 times a day,
Never even having to repeat any posts,
For our creativity will have become unbounded.

The church/social hour would then be
The only time that we'd all just be sitting around,
Instead of, like all day long now
For many of the unweeded.

The rest of life would become a blur of activity,
Weeders jumping and sprouting about
In all kinds of fun and accomplishments.
No one would ever feel bored again.

The weed is the way and the light and the energy.

Weeds entwine all things.

Weeds are our friends.

Now, I can't just prove all this by words,
But if you try hard to think that
You know you're getting energy from weeds,
Then it will come to you that you will know
For sure when you try them.
Some placebo effects may occur as well.

<u>The Holy Weed Commandments</u>

Thou shalt not step on a weed.
All weeds are to be treated equally.
Thou shalt not kill weeds except to use as food.
Love thy neighbor a whole lot, unless they are a grouch,
But more if they are attractive and you are both single.
Thou shalt not covet thy neighbor's weeds.

<u>Activities and Outings</u>

Rare and exotic weed expeditions.
Weed planting/harvesting parties.
Place-a-weed-on-every-doorstep program for publicity.
Feed-the-weed product sales for advancing weed growth.
Any questions or suggestions?

< 29 >
— Observing the Robot —

First-level people have beliefs and desires,
But second-level people have beliefs
And desires about their beliefs and desires,
Becoming able spectators of themselves.

ALL TOGETHER BECOMING

A long time ago I read all of Shelley's poems,
He being a scientific romanticist known
Who plumbed the depths of mystery,
And, too, Keats and Byron, as eagerly,
They being the romantics of the earthly realm,
Along with Omar Khayyam, the Sultan's helm
A romantic scientist who invented algebra,
As well as cherishing all of nature above Allah.

Omar was as Mr. Spock' s logic
But with the glory of life added to it,
While Shelley was more of Dr. McCoy's
Excessives of emotional romantic ploys,
But, Keats and Byron were more
Of a blend, like Captain Kirk's sure
And dashing action tempered with reason,
A man for each and every season.

So, I ended up writing poems in the styles
Of Shelley's and Old Khayyàm's wiles,
The former being flowingly lyrical—
The latter twistingly epigrammatical,
Short ones at first, very precise,
But also using them as a concise
Way to whittle down entire books
To the few gems and pearls in their nooks.

So, now, after many educated years,
I still use them to boil down the idears.

< 30 >
— Smart Machines —

Although our decisions of the instant are
Fully determined, and are therefore not free,
We may happen to learn new things, and make
Choices tomorrow that we wouldn't make today.

G'DAY BIBLE STUDY CLASS,

Someone has located the Garden of Eden:
It is now underwater, unfortunately,
At the head of the Persian Gulf, near Bahrain.

It was into this gulf that
The Tigris and Euphrates rivers
Spilled their waters in antiquity.
Nearby, the Karun River
Which bears a similar name
To the Bible's Gihon River—flows southeast
Through Iran towards the Gulf.
We discovered all this by using Google Earth.

We have also located the Ark of the Covenant;
It is in a secret chamber deep within the Temple Mount,
Underneath the present-day Dome of the Rock.

It was hidden there by Jeremiah immediately before
The Neo-Babylonian destruction of the Temple in 586 B.C.
We are leaving it there since there may have been
Five more commandments than we previously thought
And we may have broken some of them.

Tonight's homework is to have a wild evening
Sitting around reading the Dead Sea Scrolls
While drinking wine. Be prepared to act them out.

I did read the Bible, as I was Catholic [until 5th grade],
and so I am referring to some of my 4th grade notes:

God, not really being everywhere,
Moves about from place to place,
Walks around in the Garden of Eden,
Comes down from Heaven
To see the Tower of Babel,
The city of Sodom, and so on...

So, God is neither everywhere nor knows everything,
Since He must come over to investigate things.

As in... God asks Adam where he had hidden himself
And asks Cain where his brother is.

Nor is God invisible, as He can be 'seen' above,
But has eyes, ears, hand, arms, fingers, and such;
However, some who see Him are ended by
"No one can see Me and live".

Moses was OK since he only saw the back of God.
Abraham, Jacob, Isaiah, Jeremiah,
And others also saw God.

Actually, nowhere in the Bible
Does it say that
God knows everything.

I learned all these facts
At St. Bernadine Catholic Grammar School,
In Forest Park, Illinois,
Which is next to the Atomic Fireball Factory
That burned down once... but that's another story.

After my conversion to normalcy in 5th grade,
But before falling in love with my nun in 6th grade
[Another story],
I looked even deeper into the beauty
And the strangeness of the Bible
Since I was bored in school, and noted that:

Many Bible stories were recorded in writing
For the first time [they were oral before]
Long after the historical events described,
Thus creating a further history altered by hindsight,
It being shaped by the intervening events.

For example,
The destruction of Solomon's temple
Is foretold in the books of prophecy
Written long after the event,
Foretelling what had already happened...

Same for the New Testament,
But only a few generations or so afterward...

(I also found some notes from Molly McGuire,
But that is another story.
By the way, we raided the dumpsters
Of the Atomic Fireball Factory
And filled the empty desks with fireballs.)

Unfortunately,
My 6th grade nun ran off with our priest.
I didn't even know that little old me might
Have had a chance with her...
(I was afraid to ask to walk her home and all that.)

The original text of what was to become
One of the Bibles that we might own today
Was actually translated numerous times,
With each new generation imposing
Its own political and religious agenda on it.
I had a Greek Septuagint version once.
— 8th grade notes

In 7th grade, they had separated
The girls from the boys—and so we all
Just got all the hotter for each other,
Then meeting after school and...
But that's another story.

So, my notes say that Exodus
Looked somewhat suspicious:
600,000 men, along with women,
Livestock, and children,
Wandering around for 40 years
In an arid wasteland,
Just because Moses, being a man,
Wouldn't ask for directions.

Also, there was no archaeological trace,
So probably it was just a small thing
That got way exaggerated.

As for a conquest of Canaan,
It already being full
Of the original Israelite conquerors,
It was really like
"We have met the Canaanites
And they are us".

As for David writing so many psalms,
He didn't really, for the Hebrew word for 'of'
Really meant 'for', as in "for David".

Please don't be late for class tomorrow.

< 31 >

— Healthy Point of View —

Everyone's wired differently, having
Their own private, but valid, perceptions
Due to genes, learning, and mental health—so,
Each person is 'right' to do what s/he does.

< 32 >

— Fixed Personalities —

Yes, robots we are, so, therefore, major
Changes, like intro- to extroversion—
Probably even hard wired—get harder
As one grows older, perhaps impossible.

< 33 >

— Most Disagreements Eliminated! —

Since personality is mostly genes,
And the rest is hard-wired learned behavior,
We can't expect to change others, big time,
So don't bother to; let's just avoid them.

MORE PSEUDO BIBLE LESSONS

Were the Homo series of near-men and proto-humans
Merely a lucky result of the extinction of the dinosaurs
And 90% of all species by asteroids or some such?

Or did God send the asteroids?
If He could send plagues of locusts then... why not.

For those who feel that evil sprits still tempt us into sin,
We could say that the Devil's method
Is to unbalance one's brain chemicals. Not fair?
Well, don't expect 'fair' from a Devil.

Religion can always adapt to new information,
As it did with evolution and the asteroids above.

So, the infant species of Homo Sapiens continues forward,
Some ahead of their time and some way behind.
It is not totally unexpected that many will arrive at evil
And that saying they shouldn't won't change this.

Some will obsess on winning the Olympics
Or being great at something
(Doing little else along the way);
Some will go into depressions
And do bad things;
Some will get so anxious that they will
Hit their children and spouses,
And many will slip into
The sit-com life of being selfish.
This is human nature as it really is.

Making Adam, sending Moses,
Restarting with Noah, and sending Jesus
All did nothing because it was all a myth.

Science, of course, continues
Inventing helpful things;
Schools continue new programs like
Good For the Sake of Good (GSG),
By, say, putting drunk driver's wrecked cars

On the school lawn by the entrance,
Asking for compassion, and so forth,
As in Rachel's challenge.

I suggest even more focus
For the young and impressionable.
There could be a class
In which students daily log
How their thinking out of consequences
Helped lessen their problems,
Or, when they merely reacted without thinking,
How problems arose and bloomed out of control.

Such thinking ahead might then
Become as routine as doing math calculations.
Why not focus a lot on the actual living of life?

We have, perhaps, zillions of years to improve,
Which, at the current rate, may all be necessary,
For human nature is now what it has become.

All could very well go the other direction as well.
Overall to date, is the human race progressing?
If not, why does the same evil stuff keep befalling it?
Is the gene pool degrading? Why are the prisons full?

Will drug-users, abusers, gamblers, sports nuts,
Gang members, workaholics, and all such
(The list is too long)
Reproduce so much more of the same
Until goodness becomes a rarity?

Still, what an adventure it is to be alive
At this time on this pale blue dot
In the middle of nowhere
Between the eternities of forever—
As it was and ever will be...

Evolution via natural selection
Endowed human mammals with the notion
Of looking for Intent in nature,
And that helped us out a lot with our real environment.

That it became somewhat innate is The Problem
In as what people still make up about the Intent.

It's still that once one uses a specific word,
One has to declare the word in all its specifics.

I know we all have a vague idea of what God means,
But ask ten people ten questions about God
And one will find that there are ten ideas
About god out there, not just one.

If one uses a word, one must fully explain that word,
And in light of the Theory of Everything,
One must also do that on
Scientifically satisfying grounds.

Many have said
Semantically impossible things
In which the words sound good,
But make no sense, like, say,
God is the Universe; but, like a rose,
It's still the universe by any other name.

Many even said what God is not,
Such as being undefinable,
But that only supports the claim
That to say "God" says nothing,
For it indeed defines nothing.

Everyone thinks that God knows everything,
Can do anything, and is everywhere,
But really, actually, for sure and with no doubt,
Poor old God was just an advanced alien bumbling along
Through some carbon-based experiments
As a student of biology...

In retrospect we can see that
Any higher composite Mind
Would have to be upwards of our evolution,
Not something tiny, simple,
And non-compositely fundamental.

He's not really God, but we'll still call Him that
Since He created us in a lab experiment.

First, He threw out some debris
That became our universe,
Since He really couldn't make
Any original stuff,
And, being limited to that,
Had to wait
For 13.75 billions years
For homo sapiens to appear,
Then tampered with our DNA
To put the final perfect touches
Upon our human nature,
Making "Adam and Eve" in a Garden.

Well, good old God,
Not really knowing didly-squat,
Was mighty darn surprised
When his human nature design
Immediately flubbed when they ate
The apple forbidden to them.

"Jeeese", thought God,
"You tell them not to touch something
And that's the first thing they do."

So He threw them out of the Garden
To see if that would help.
It didn't, and so God was very surprised,
But all were fruitful and multiplied into the millions.

God waited around,
Thinking that surely some more evolution
Of His new DNA masterpieces would do the trick.

It didn't, surprise, surprise,
So God found the best man on earth, Noah,
Saved him and his family
And killed all the rest of humankind.
Things would surely improve now,
For this was like breeding.

Nothing improved whatsoever,
Even more folly and wars going on,
So, a much more surprised God
Sent Moses down with the Ten Commandments.
This would help change the masses.

Well, things did change; they got worse,
And so God, shocked at this turn of events,
Sent many plagues of locusts
To scare the people into shaping up.

This did not work and
So God was utterly astounded,
So God sent some prophets,
But nothing much changed.
"Darn," said God.

God then sent Jesus to preach goodness,
But they crucified him.

Wars, stealing, murdering, plundering,
And name calling continued unabated,
Different religions even warring against each other.
This was all really getting out of hand.

God sent even more prophets, such as "Bab".
No effect. Shock and surprise.

Earth's problems got worse;
The Nazis almost conquered the world.
11 million died in camps.
God, of course, was limited, apparently,
And could do nothing to help.

Finally, God, realizing that
His experiment was hopeless,
Turned in His lab report
And soon flunked the course.

(This is a true story handed down
From some ancient historians
Who knew everything.)

The phenomenon of reliably consistent creation
By causal intelligence lying behind it
Is philosophically and logically impossible
Without more causal intelligence lying behind it, etc.,
That is, a system of intelligent mind is a system,
Having parts beneath that are more fundamental
Than the resultant system.

Where does it end (begin)?
It cannot be with mind,
For mind is composite;
The regress must end.

I was at a wedding in Canaan one day,
Drinking water, well, like water,
When, all of the sudden, it turned into wine.
Well, the problem was that
I just kept on drinking it like water...

< 34 >

— The Information Age —

More and more the myths of past ignorances
Give way to the solutions of science,
For example, a person's life force and
Basic traits are her genetic blueprint.

< 35 >

— The Bad Old Days —

Life's still emotionally primitive—
Negative feedback mechanisms in
The central nervous system, now useless,
Still send thousands-of-years-old messages.

ON LOVE

Love is the finest refreshment of mortal life,
Providing as it does a glimpse into the heavenly state,
A vision which, if maintained, can last well beyond
The initial perception and for all of one's life.

So, I say that any time not spent on love
Is time squandered in absolute waste,
That if you are idling, not loving,
Or, god forbid, hating,
Then life is a-wasting;

For love is the greatest experience on earth,
And so I have often sought it out, found it, received it,
Given it, and lived it as life's one great happiness,
for there is no other joy that compares—
Love being the truth of all truths.

Who has not forgotten that first kiss
And the magic that attended it?
No one, for first love touches one deeply and forever.

People newly in love glow for weeks on end.
There is nothing like love, although, strangely,
Some do not actively seek it out,
Perhaps for fear of rejection.

But, even love's worst pain is sweeter by far
Than any other pleasure; there is, indeed, no contest—
And to love and lose is second only to loving in triumph.

Not merely just a pleasure,
Love refreshes, creates,
Invigorates, and provides sustenance
Of spirit and life itself.

Without love there is no life,
at least none worth living.
When you give up on love,
You begin to die.

Love knows no laws or restrictions,
For mutual passion is a law unto itself.
Love is the cure-all,
Both for those who receive it
And for those who give it.

The one tragedy in life is not death,
But that some people do not love—
Aye, nor do they live,
For the fear of the one
Is fear of the other.

So, by all means, if you love someone,
Go to them and tell them so right now.

It is said that the loving are the daring,
Perhaps because they seek the ultimate adventure,
Often risking all for that which lies far and above
The commonplace, that vision into paradise.

Imagination weaves a fairy tale
Of love and romance,
And the mind that is alive
Soon brings forth the phantasm into reality.

Placing our very life and happiness in another
Through love is the greatest gift one can give,
For it is the gift of oneself.

Unconditional love is a true gift,
One without strings attached,
One without any motive for gain in return.

Oh, of course, we are human and can love
For the sake of being loved in return,
And this is not in itself wrong;
But, when one loves for no other reason
Than for the sake of generosity and loving,
Then this is a saintly type of love
Which is above the others.

True love loves people for what they are;
Not for their qualities in particular,
But for the whole person.

It's not that we love someone
Because we need them—
For this is quite immature—
But that we need someone
Because we love them.

It is, you see, love that is the origin.
Love begets love and love,
And this, in turn, begets more love,
And so on, making us even more loving to others,
Until Heaven is indeed brought down to earth.
Real love is its own reward.

Identity is not lost in love,
For true lovers do not sit
Looking only into each other's heart,
But, rather, look outward,
Both in the same direction.
It is a seeming violation of arithmetic
That in love two become
Much greater than one plus one;
And that the two, nevertheless,
Do not become one,
But remain as two,
Yet still share the same vibration
In their souls.

It also seems to be a paradox that love,
When divided, is not at all diminished,
But that each individual love
Multiplies to exceed the lot.

One can never run out of love!
It is a miser, indeed,
Who withholds love from
A capacity that is boundless.
Hoard not that which can be given.
Give love, and even more love comes back full circle to you.

What a joy is it to experience life's wonders
With someone you love—
Oh, walks, and plays, and dinners
Are great enough pleasures when taken alone,
But note how much better they become
When you have someone to share them with.

Another bonus of love is, that,
With it behind your actions,
You may soon find yourself doing the impossible—
As love's inspiration carries you along
Through any kind of difficulty.

For me it was an inspiration to write.

Love and a kind heart are much alike,
And one is equivalent to the other,
Love being a triumvirate of truth,
Beauty, and goodness
Blended into one great purity.

We do not merely love—we are love!
We do not create—we are creation itself.
We don't just live — we are life!

There are many forms and faces of love,
Such as brotherly, sisterly, motherly, fatherly,
Romantic, spiritual, professional, and physical—
And it often depends much upon the circumstance
Which one is the most appropriate form
To give to a particular person.

PASSION & SENSUALITY

Reason spoke to Passion, with logic cool:
Quench thy inner fire, lest it burn us, fool.
Said Passion: I know WHAT I feel, not WHY;
'Tis better you take heed of me—I rule!

I give no reason for love's passion planned,
Because to do so would be secondhand;
For the Heart and Soul have many reasons
That Reason could never understand.

Convince me, Nature, that Reason is right,
That the strength of the heart is not in flight.
I'll plunge into the depths of thought and love
And tell the spirit I'll defy the blight.

There's an urge between root and flower,
Plant and soil, leaf and sun, air and water,
Daystar and planet, valley and mountain,
Wind and mist, man and woman—for ever.

Since we're embodied, we have desires.
Suppression of desire strangely backfires,
Since—and this is the paradox—it takes
A strong desire to overcome desires!

Sensual bliss should not be a lost art;
For the body is an integral part
Of the human and joined with the spirit—
Realize yourself with whole body and heart.

It's unnatural to suppress a natural urge,
For this is distortion—the most unhealthy purge!
Lack of food, sleep, or sex can lead to neurosis,
So—let natural functions freely emerge.

Head, heart, body, and soul were together built—
So why separate them? Merge them, so thou wilt
Have more awareness of life's experience
And free sensual joy from feelings of guilt.

In the Eastern world, lovemaking is an art form
In which body and soul in unity perform.
The Western approach is by joyless guilt deformed,
Though sexual energy is a human norm.

Hindu goddesses aren't virgins thought of—
Their healthy desires are free to rove.

Enlightenment is sought and reached through the
Profound experience of sensual love.

Your pleasure depends on the permission
Of others if you abide by the shunned
Taboos of society, parents, or peers;
So, only you need approve the mission.

No matter how ethereal love's spirit,
Sexual union is still requisite,
Because we are physical beings— and since
No union is meaningful without it.

Pleasant smelling scents lift your heart and mine:
Essence of lotus, rose, amber, jasmine,
Night-queen, myrtle, saffron, and sandalwood
Stimulate the inner spirits sublime.

The tulip lifts her blushing cheeks to me,
As wandering winds caress the rose tree.
She wears a spring smile and pours dewy tea.
Yes, I'll drink you long and deep into me.

True kisses are always new; they never
Lose their freshness; for, like falling water
Or the cyclic moon, the power of love
Renews itself to sustain for ever.

So much sweeter sounds are your lover's sighs
Than the groan of a war that wins great prize.
Just one taste of true love by far out buys
A Sultan's wealth in some rich paradise.

The rose is the flower that the bee cruises,
Meeting there the butterfly that love chooses;
They unfold the petals of the blossom,
And drink the nectar of love's sweet juices.

Her scent was ripe and her name meant nectar.
Exotically blossoming I found her,
And I buzzed my way into her flower,
For I was the bee and s/he my partner.

I drink her wine into my two-lip cup,
As meanwhile, my giver-of-life comes up.
Petal by petal, her rose wide unfolds.
Passions grow from the dew on which we sup.

Like water, a woman is slow to boil,
And, likewise, slow to cool down afterward;
Man, like fire, can be ignited and quenched;
But fire and water in balance make steam!

We rarely sit in front of fires created;
So, the light of sacred fire has retreated
Into our subconscious; thus, candle flames are
Harmonious with love's passion stimulated.

Fire inspires everyone under the sun,
Especially lovers having their fun.
Firelight flickers, playing on nakedness—
The inner fire being as sacred as the outer one.

Spontaneous, endearing acts allow
Dead yesterday and unborn tomorrow
To be dropped from the calendar, for,
In the love-temple there is only NOW!

Our passions smoldered, like incense fuming,
And brightly burned, the candle flames luming,
Waxing full as we consumed the body,
And then rose as spirit smoke, mushrooming.

Your wine, my persona radiata,
Fills the golden chalice—oh, Sultana,
I'm intoxicated by your love-stream
Flowing freely—oh dear, amorata!

You enclose my universe, yet it's boundless.
You fill up my universe, never the less.
I'll fill up your emptiness with my fullness.
I'll empty your fullness with my emptiness.

My dearest: Your wet lips' sensual pout
Draws me to your flaming well, in and out;

Love's sensation touches us everywhere—
At last your sweetwater puts my fire out.

Hardness rises from the earth element.
Secretions flow as water's element.
Sexual friction evokes sacred fire.
Ecstatic pleasure fills the firmament.

Oh never did I hear a sound so sweet
As when you moaned like a panther in heat.
You took me on a wild jungle ride,
Then purred like a pussycat at my feet.

Let the fruits of your lovemaking ripen,
By remaining in close union, so then
Energy and spirit can be absorbed...
You'll blend in a mystical way—Amen.

WHAT LOVE'S MADE OF

Some may ask of Life: *How does one find love?*
Life says, *Be still! Don't look far or above;*
Stop—let love's butterfly alight on you,
For that's the touch that romance is made of.

Love lightens life's cross of heaviness,
Its dreams coloring the drowsy darkness.
Love's subtle hues paint life's panorama,
For it's the bond that secures completeness.

Some can't seem to give a love that's fair,
Or won't—since they don't even care to share,
Or, worse yet, they waste love by hoarding it;
So therefore I must give more than my share.

Let me give all the love that ever was
Of such and more I have been dreaming of.
I hear the call from within and above:
To live this life for the purpose of love.

I mend broken hearts with a love that's real,
Drying the tears of sadness that congeal.
I weave living dreams out of fantasies,
For I believe that life should be ideal.

Might I be your angel, enfolding you,
Shining brightness upon you, holding you,
Nurturing, and carrying you aloft,
Where you belong, dear, far above it all?

Life suddenly fits me like a glove,
As I float on feelings like a dove,
Renewed energy giving a shove.
Well, could it be that I am in love?

Summer follows us around, elfin queen
Of hearts, as we lay snug in winter green,
Your glowing pixie crown lighting the scene,
Your curves spooning, the ears pointing away.

Strong, in the heart of the working day,
My love for you burns the hours away;
Long, in the soul of sleep's deepest night,
I'm with you like a dream that will stay.

Like living lenses, we mirror our love;
In feedback loops images spiral above,
Echoing as infinite reflections
Which fill up the scene—that's what love's made of!

LOVE

Men and women cannot exist in isolation,
For much like valley giving rise to mountain,
The nature of one makes necessary the other—
When they're joined in love, there's wholeness again.

To find yourself, lose yourself in another;
For s/he will touch your being and therein share,
Gently unveiling your heart, soul, mind, and sense
Till there's no place to hide! You're found, forever!

If love were easy to find, have, or say,
Then its meaning's worth would soon fly away;
But from steady effort love grows to lead
Yearning to fulfillment—and there to stay.

Love is the mutual creation of identity.
To be in love is NOT a loss of independence,
But rather a shared identity with the lover
That does not destroy the identity of the other.

Love is GIVING without gain in return;
TAKING is selfish—will we never learn?
Graciously accept all that you receive,
And give kindness to everyone in turn.

Love matures when lovers let it flow beyond—
Free to wend its way to places dear and fond.
Love's butterfly prospers when winds blow free;
Unconditional love never binds—it bonds.

Freely given love returns on the wing,
But if you KEEP your love you'll have nothing;
It's a most wonderful paradox—
GIVE your love and you will have everything.

Arithmetic theory fails in love's plot:
Love when divided diminishes not,
As would else we know, and vanishes not—
Each love multiplies to exceed the lot!

The capacity for love is boundless—
No piece for us and fraction for the rest.
Since the sum of love's parts exceeds the whole,
We can give and give love, never-the-less!

Poets translate what's within and above,
To exhibit truths from depths unheard of.
There's one deep truth that I know to be true;
I'll tell you too: The truth of truths is LOVE.

The meaning of love is in its GIVING
When there's no motive towards obtaining;

TAKING is the opposite of giving!
Caring? Sharing? They're reasons for loving.

Of a love-sweet companion take your sup,
While s/he as your chalice is lifted up.
Drink deep the wine that satisfies love's thirst;
Drink up—before winds of change dry the cup.

To your lovers all your kisses bestow
When life's colors glow in your rainbow;
For as long as love's kisses can live,
Neither age nor time on your life will show.

LOVING is what this life is all about
To give and have it is to live all-out!
Love's the finest thing! Can you do without?
Then, why, oh why do you not seek it out?

Small town summer, picket fence, grandmothers
Playing cards on the front porch—Hearts lovers;
Country girl brings cookies, rhythm flutters;
Smiling, all approve of what love discovers.

As I wandered along the romantic way
With one who would drink life's sadness away,
I realized that the cost of a loveless life
Was much too high of a price for me to pay.

We sent out emanations of love fair
That were sweet, soft, and smiling on the air—
A scented mist of liquid love that filled
The scene with its well-being everywhere.

S/he is the elixir that fills my cup,
The perfume on the breeze that lifts me up
S/he's love's essence distilled into being,
The passion-spirit that opens me up.

Driven not by desperation or pain,
But purely by love alone, I sustain
Affection through the goodness of giving,
For true love's but pure love preordained.

A mutual self we form, one both friend and lover,
Touching soul-to-soul by language we discover,
Opening each other up to connect our selves
Now we TWO total more than ONE + ONE other.

Initial fusion requires heat and excitement.
A steady flame then insures passion's maintainment
And continues the honeymoon's enchantment
By sustaining a reaction that's permanent.

Beautiful sentiments call attention
Style/substance = emotion/motion.
Gather sentiments, place them on the scale
They weigh much less than one lovely action!

I'm painting a picture for you to see
The vision of romantic destiny.
Like ghosts, we emerge from the photo film,
Virtually into reality.

Speak freely to me—let me hear your song,
For what you know and feel cannot be wrong;
It is genuine because it is yours.
Such intimacy does our friendship prolong.

Flowers grow from their many roots, upcast
Some have passed, some are steadfast. The contrast:
Those which grow much too quickly, wither fast;
Those which grow steadily and slowly, last.

In the soil that we shared, these plants we chose—
Truth: tulip, goodness: lily, beauty: rose.
Nurtured with life's care, they wave to and fro;
Storms can't scatter the flowers that love grows.

I've never, in either life or dream, found
Anyone as alive as you—you, who 'round
Me as a living presence glows and gives
Me youth, life, love, and a spirit newfound.

We rose higher and higher, past cloud nine,
Through seventh Heaven, to a golden shrine

Of Love where few have ever entwined,
For we let love build but never decline.

Senses melt away, drip by drop by drip.
Impressions flood the speechless spirit.
Emotions flow free for the heart to read.
Love has drawn me in—I'm dissolved in it.

Your spirit calls, steam risen from the rain,
A missing so sweet that it's almost pain;
The future's heavy, swelling with promise
Of the season when love can breath again.

I never knew that love could be like this,
A wonderland of peace, joy, and bliss.
No, I'd never known where I'd never been,
That such a world could be found in a kiss.

Together we sing in a fugal voice;
For we live in two-part harmonic choice:
We're opposite twins in love, a canon
Chime in which we in unison rejoice.

Like voices merged in the Canon Pachelbel,
We speak as one, as the knell to the bell,
S/he saying what I thought and vice-versa,
In tune, in unison, yet parallel.
Fugal voices blended, parted, and long
Wove in and out, the music sweeping strong,
And onward, upward, inward, and outward
Till being was left to the spirit's song.

As I love and am loved in completeness,
Then this world, with all of its foolishness,
Work, hurry and scurry, pain and worry,
Does fast fade away into nothingness.

Cares floated out on the tide, and then some;
Sun-sparkles glimmered, danced, and swum,
Alighting on my mind, there to become
Ideas about the loving night to come.

Your partner's heart beats dear against thy own
Where you're safe, warm, and completely at home.
Surrounding the blossom of your flower,
S/he enrapts you like the words of a poem.

Thy heart touches my own; no, 'tis more I love thee!
Yes, much more art thou loved, the me's now in thee.
Thou art the soul of my soul and mine is of thine;
Nay, 'tis more than that: thou art me and I am thee!

Can I ever fathom the source of love?
Perhaps its fount springs from Heaven above?
Just this I know: Love's rhythm resonates
Beneath words and thoughts, in depths unheard of.

I am immersed in love's boundless dream,
Floating in peace on beauty's quiet stream.
Truth is now clearly seen, so bright and right
Purity's goodness swells each sparkling gleam.

With sparks from passion's smoldering embers,
We ignite from all that love remembers,
And steam through emotion's ocean in our
Relation Ship—of which we're crew members.

There on some remoter shore of human soul
To which I helped restore life and spirit,
I learned that love was the only flame that lit
This life—for she had taught me how to give it.

Awash on a love-made shore, we overcame
Our senses, leaving them behind, unclaimed.
As we floated free, quenched in sunset sea,
Basking in reflections of the scarlet flame.

When the sun burns up, and long after
The Earth grows cold from that disaster,
When galaxies die and rotate no more,
Then what remains is our love thereafter.

In love relationships, not only banish
Criticism, nagging, name calling, anger,

Punishments, and yelling, but replace it
With encouragement, support, and caring.

(The Trouble with Love)

Only a few words rhyme with the above,
Like the overflown dove, the heartless shove,
And the ill-fitting glove. Alas, love's rhymes
Remain unheard of, or aren't well thought of.

LOVE RECOVERED

A puzzle, if one muddles, can be made to fit,
The parts making a seamed whole, bit by bit by bit;
However, people in unions have edges that sit
Perfectly on one side but not on another.

Loss is painful when leaves fall, but you cope;
As always, new attachments form with hope.
The cycle of the seasons mirrors all—
Life is a generous kaleidoscope!

Love is the ultimate reason to live.
To for-get, it's necessary to for-give.
Habit bows to originality;
Emotion's energy becomes motive.

Can one really realize life's benefit,
And live every precious minute of it?
And can such awareness withstand the strife?
Yes, if you're a lover or a poet.

Children back to school, Autumn in the air,
Avenues quiet, a bedroom upstairs,
The warmth of a nook; we caress up there
From flesh to spirit, loving with great care.

Soft breezes blow, caressing me and you,
As we kiss the roses and drink their dew.
Reason and passion soon merge into one,
As truth and beauty make their rendezvous.

CONSCIOUSNESS

Perhaps information from the brain
That gets globally bound to
"Surface on the mind"
Becomes the 'reality'—
The consciousness observation.

This is kind of like the "It from bit"
Of the quantum realm in which
The bit (the information) becomes the 'it'
(The collapsed state of the actual,
Realism having become),
Although the quantum has
No pre-existing properties,
Making the outcome random.

The brain has a pre-existing state
That must correlate with
What's witnessed in consciousness.

So, consciousness and what's in it
Somehow becomes from
The information in the brain, that is,
What thoughts, actions, and feelings
That are of the instant,
This all then repeating on and on,
The consciousness states, too,
Becoming part of our repertoire for future use.

< 36 >
— The Fall of Evil —

Emotions are but molecular events,
Some forced upon us all, like jealousy,
And some others, like aggression, born from
Low serotonin, NOT from the Devil.

FROM "THE LAST KNIGHT'S ALMANAC"

The clear skies uncovered a hidden wonder—
The pale Unicorn full moon just yonder
That marked another month gone by—
There, plain as day, in the twilight sky.

Memories danced again to yet another time,
As All-Man's-Knack made reason and rhyme.

Thus, the day of sunshine and moon glow.
Yes, the Lion-sun rules the toiling day,
But the Unicorn-moon, shining bright,
Brings wisdom and calm in the night.

At the magic times of dawn and twilight,
When the lion and unicorn battle together
For the heavens, only beauty and goodness
Can ensure the victory of wisdom.

After twilight had gone, the October moon
Of the Harvest feast rose up in the east,
Looking much larger than it really was.

She didn't feel so alone as she had thought,
With that man-in-the-moon she sought,
Looking over her shoulder.

She felt somehow protected by that lonely sky face.
All was so bright that the only darkness
About her was her shadow.

She entertained the thought that there
Were but three at the camp tonight:
Herself, the moon, and her shadow—
A thought that everyone someday
Entertains on the most moonlit of nights.

The moon, which often seemed cold-hearted,
Was warm tonight. Wherever he was that night,
She knew that he would see it too. She drank the
Last wine of the Grail, now a part of the legend.

Some two centuries after the time of Alexis,
Li-Po, of China, would write,
"I take a bottle of wine and I go drink it
Among the flowers. We are always three—
Counting my shadow and my friend
The shimmering moon...

Happily, the moon knows nothing of drinking,
And my shadow is never thirsty.
When I sing, the moon listens to me in silence.
When I dance, my shadow dances too.
After all festivities the guests must depart;
This sadness I do not know.
For when I go home, the moon goes with me
And my shadow follows me."

That night he observed the Harvest moon
In the sky, and in it saw a hybrid face
Containing the features of Arthur, old Arthur.
It was said that when Arthur died,
Merlyn rearranged the mountains of the moon
To resemble Arthur's face—not as he now appeared,
But as how he might have appeared
Had he ever gotten old.

He arrived; they embraced and kissed,
And then they raised their glasses to the sky.
During the toast, a flock of ducks flew high,
Barely visible except when they crossed the moon.

"A toast." Percevale declared,
As the jewels of Orion's belt twinkled,
And as the starlight passed through their glasses;
Perhaps those three stars were the twinkling
Of old Merlyn's three eyes, those which could see
The present, past, and future in only one glimpse.

"May we all toast again in three places
Of the Universe," said Percevale:
"Here, where we stand now,
Then in the old hall at Camelot,
And, finally, in Heaven."

I have never seen a bluer sky
Than that of October's on high.
Perhaps it's due to the cool dry air.
The vision is enhanced by the foreground
Of the colorful orange tree leaves.

Following the harvest, the moon that yet
Shone was a strange sight in its late morn-set,
A large chunk was missing from its battered orb.
There was the sun well risen in the east
Seeming to balance this moon of the west.

They made our way through a lonely upland
Wild and still, where October's last wind
Whispered at will for the souls of the dead.

Towards evening on first November's day,
The first quarter moon rose later this day,
Sitting atop the evening star to pray,
And then rose later and later each day,
Drawing away from Venus, going away,
Thereby adding light in its own way.

This is the last true blue
That we shall see for some time;
It's only fitting that it be the best of times,
The bluest of times.

No leaves, no warmth, no sky,
No snow—November.

November is a most difficult time.
The glory of the summer and of the leaves is gone;
It seems like it has been gone for years.
The spirit of the holiday season is not yet at hand.

The grey and rainy skies are a stark contrast
To the dry blue skies of October.
There is no snow yet for winter sports
And the land remains barren;
The land is dead, and the very year itself
Continues to die in the night.

The day is so short that
When one gets home for dinner
It already seems time for bed.
Time for hibernation perhaps.

To these feelings we add the specter
Of a long, drawn-out winter.
Now we even long for February.

Come December, we'll await, when auroras
Will set fire to the polar heavens
To give some color to our lives
During the festival of the Yule.

'Second' summer was brief this year
And some weeks passed without a tear.
Chill winds now hastened our approach
To the nearest inn, as we beheld, by coach,
The rising omen of winter in the late night sky:

"Orion, King of the bejeweled winter sky,
Backbone of our frozen nights," said I,
"Wield your sword ever above our headen bow,
But please, for our knightly sake, never below!"

There it was, smack in the road a head—
A huge yellow beast rising dead ahead,
It growing larger as we the seconds read!

Now right in front of our eyes did it lie!
It was, of course, the moon not yet high.
"The November Frost moon is even more
Impressive than that of the Harvest lore,
For it is so colorful and intimidating."

It rose straight into a thunderhead's din
As an old lady opened the door of the inn.
"Tit for tat," said she, as she farted at the thunder.

The first of December was clear, with rime,
And Venus could be seen in the daytime—
It seemed to be directly in the swoon

Of the early setting first quarter moon,
And was even brighter than that moon.

Sure enough, the time came when that moon
Gave birth to Venus after its eclipse, so soon,
And they both together set from the skies,
A wonder of wonders to my open eyes.

When the full and gleaming December moon
Shone on the snow,
It made for the season's brightest night,
Brighter than the most dismal day.
No one saw this Cold, Yule moon, however, but one
Sir Arthur, once King,
And may he ever watch over us as we sleep.
This was their prayer that night.

It was but one of the few full moons missed
By Percevale's keen eye, but he dreamt of it,
As did many others that night when the moon's tides
Tossed them about on the swells
Of resurrected memories from forgotten seas.

They woke up suddenly with a start.
Even before they were fully alert
They realized that something had changed.
They couldn't quite name it, but something...

It was very quiet outside, and this silence
Was somehow different, for noises were muffled.
The daylight coming in the window was brighter,
Barer than yesterday's.

Then they realized that the season's
First big snow had fallen during the night.
The air, the trees, and the inn were hushed
In the snow. So, now it was winter.

Their memories swelled for a while.
To the changes in the weather, in the seasons,

They thought, we are sensitive,
Despite our civilized attempts
To remove ourselves from its raw influence.

We never truly sleep. We are animals yet.
The first snowfall you won't hear, or often see,
But its advent will bring you from the deepest sleep.

December's foggy freeze descended upon the inn
And the cold air seemed to crinkle about the body.

Guinevere had seen many such winters,
Some from cold and lonely towers,
But winter's embrace always seems to surprise us.

This day, many a gravedigger cursed his task
As he shoveled at the snowy hardened ground
In an attempt to bury the Vikings killed last night. The
clanking could be heard throughout the day,
And the bodies were as stiff as the ground
That yawned to receive them.

Christmas came and went at the inn,
Hardly making a dent in the festive atmosphere;
however, the stone walls and withered gates
Of the inn warmed with the festival of the Yule,
The bakers' cakes, the rituals of the Druids,
And the cutting of the sacred mistletoe
From the chosen oak.

Lazy winter days turned into weeks.
Percevale began to daydream of the Tropics:
"The moon is growing larger, towards first quarter.
What a strange sight is the tropical moon—
It fills up with light from the bottom,
Not from the side—but that's Equatoria!

Proud Orion the hunter and his three hunting dogs
Stride high, right down the center of the sky,
In Equatoria, for that is where you will find
The celestial equator there—directly above
The earth's equator, not near the horizon.

Alliances of the north are broken now, here,
Neither north nor south of the moon's path.
'I am the only white face, all others are brown.
And my name is Taliesin, the King's poet.
You dream of me, Percevale, for I'll soon return.'"

On January 4, the earth was at perihelion,
Its closest point to the sun. It was a new year.
The full Wolf moon rode high in the sky
That night to guide all on their way.

When the dead Orion is seen no more
And the Archer of the sky pursues
The Scorpion to his death to avenge
The poisoning of that great hunter,
Then it shall be the spring of wonder;

That will be the time of the greatest glory—
Once Mars the warrior catches the two giants.
This will carry us well through the time
When Orion again rises once again
To pursue the seven sisters,
The seven daughters of Atlas.

"Perhaps it is the blessings of the gods
That we see up above," she offered.
"How nice of them to provide dotted pictures
For us in the form of the sky's constellations.
How clearly they can be seen now
As the new crescent moon sets.
There is the old moon in its arms,
Like our old life as it rests
In the arms of our new one."

Delicate stars drifted to ground, as snowflakes,
And the day was filled with snow-bow wakes
As the morning wind stirred up the stardust,
Old and new. Hibernation soon overtook all
As the full dead-winter Snow moon
Brightened the frozen night into noon.

Meanwhile, the Big Dipper had nigh

Swing round in the winter sky,
Looking like a gigantic question mark.
Later, showers of ice needles steadily fell,
Until the sky cleared, near morning's bell.

Then we shone in the Venus-light of morn,
Following that rising morning star just born
On through the oncoming dawn
And into the daylight, for it was shining
Brightly in the night,
And casting our shadows on the snow.

"What is on the dark side of the moon?"
"We don't know, Percevale;
Mankind is still attached to the earth.
While there are some now who fly balloons
Of hot air, they can hardly go high enough
To investigate the mysteries
Of the moon's dark side."

"Indeed we all have dark sides.
I have learned to live with mine."

The great astronomers of the Land
Had finally reached the King's court,
With only two days to spare.

"My King," they cried in unison
And out of tune at that,
"In two days hence all of the planets
Will be in as straight of a line from Earth
As is ever possible! If this were not enough,
The full Worm moon rises as well that day!
This alignment dooms us all."

Surprisingly, on Doomsday Eve,
The kingdom did not erupt into madness.
There were no robberies,
For what good would wealth be after Doomsday?
There were no murders,
For who would wish to blacken their record
So close to the Judgment Day?

There wouldn't be time enough to repent!

No, it was only the best of times.
People warmly thanked their friends and enemies
For the spirit that they had shown,
Explaining that they were always too busy
To do this before. All was tranquil.

"See there, Percevale," said Taliesin,
"See the mist about the planets and the moon,
Especially near the Sea of Vapors?
The opiate rim vapors of all the great planets
Have calmed the people.

These lofty visions of the great unknown
Have induced the people to forget their day-to-day
Minor grudges and realize the goodness of life—
To realize, almost too late, what could have been.

No Doomsday is this;
It is the first day of our new world!"

{Goodbye Moon, Hello West Wind}

Yes, they were tired, unlike our Lion escort,
As they finished a March of thirty-one days.
There they all stood, like fools, looking at the moon.

The moon just sat there and waited
For someone to speak. No one did.
Finally the jester said, "It's not so funny—
The weather we're having, I mean."

"Yes, the days seem to be getting longer,"
Said the moon. "Soon I and my friends
Of the night sky will be all but obliterated
By the long days and the lingering twilight
Of the British summer. I trust that you
Will not forget your old pals like me
Who carried you through the winter!
We trod the snow with you—remember?
I was your shadow; my footprints followed yours."

"I will never forget you, old man," said Taliesin.
"We will stand by your coppery side
Even in the month of June as you disappear
During the short mystic night,
The night of the broken moon—
When you, our pale-faced King of the night,
Are disrobed of all your borrowed light
By an eclipse.

But, now you must surely realize
That the balance tips
Between the day and the night;
We have the equinox
And the spring to consider!
Now men's hearts must turn away,
For a time, from the fathomless sky
To the outdoor events
That do unfold all around us.
How we love to be taken ill
With the Fever of the Spring.
Like the April fools that we are,
We fall for it every time.

Inebriated are we all with
That first breath of spring.
Not just a breath, mind you,
But the beloved zephyr,
That lukish wind from the west,
Never a wicked wind,
But a lovely soft wind
That caresses one's body
From head to toe
As long as you will allow it."

"Now I remember the spring," related Guinevere.
"Now, this morning, as we wash our faces
In the dew, I remember! And now I know,
And how do I know—that life is good,
Especially good, when the west wind
Softly evaporates the morning dew
From our bright faces."

"Indeed," replied Percevale,
"I am always amazed, year in and year out,
When, in spring, nature reinvents the world."

The moon never could get used to being ignored
In the early spring when men's eyes turned
Ever downward to the up and coming crocus.

So, the moon burst forth with many tales.
First, the moon sliced through the clouds
And cut them to shreds.
Next, the clouds in the other half of the sky
Went black, even with the sun behind then
And the morning bells began to chime.

And on a new April day,
Snow fell from the midst of the moon—
flakes like fine willow flowers,
Like shreds of silk torn from the clouds.

The old moon man blew it this way and that
Until one could hardly see from here to there.

"O moon," said Percevale,
"How little faith you have in us!
Did you really think that we could forget you?

"Taliesin, my poet", asked Percevale.
"What words do you have for us
On such a snowy April day?"

"Well, my King, I say to you that April
Has long been classified as portion of winter,
Not of spring, and that an April blizzard
Is not unexpected; furthermore, the winter,
Like death, seizes on men whether prepared or not;
But that May, like the virgin of Virgo,
Will quickly work her spring charm
Into the embrace of winter's icy arms
Such that all shall be warm and flowery for us.

Even now the wing'd musicians do sing

To entertain the bashful spring.
Now, sit down and hear the snow melting
Under Nature's eye in the sky.

Put your drink in the shade
So it will stay cold,
For the bright light melts the cold ice.
And listen:
The rivulets run down to the shore;
The trickling is heard by us
As we are laid sunning in the snow.
Great white beasts lumber through the sky
As clouds and then disappear at twilight.

OK, moon—
You full and budding Pink moon,
I have a question for you.

With Percevale and Galan no longer knights,
But Kings, who will be the last knight now?
Who is the last knight of them all?"

Replied the moon, "thanks for the easy question.
Don't you remember? Why—I wear his face!
It is Sir Arthur, once King, of course,
He is the last knight, watching over you
From my visage as you can but smell
The fresh-turned-earth-mold!
Now chose your mate,
The joyful spring to celebrate.

...

The Viking ships sailed underneath the cover
Of the clouds which hid the full Strawberry moon. Soon the
first of many river villages began to fall.
The first village took the brunt of the attack,
And the rest were empty by the time
That the Norse men reached them.
An ugly infection was quickly spreading
Through the main artery of Britain.

Alexis eased the old ship past the Vikings
Sleeping on the shore, her navigation aided

By the full moon, now showing,
And by Percevale's river knowledge.
"They don't see us yet, Percevale."

The monthly tide wave had already entered
The back part of the river near the harbor
Where the young knights dined
At The Port of Missing Men,
Where the warning bells rang and the diners
Lifted their feet, as the high deck was washed
By the splashes of the river-bore.

Of course, it would be much worse
Farther down the river—
It was a wall of water growing higher and higher
As the river channel narrowed in the Severn.

The Viking ships in the rear of the fleet—
The older, weaker ships—were the first to fall
As the river-tide bore them down—
The ships were smashed against the rocks
In this very narrow portion of the Severn
Where the tide was at its highest.

Thoralf's Viking flagship tried to beat the wave
By veering hard to starboard to gain the safety
Of a river island, but this was a bad move—
As the mighty warship was hit broadside
And was quickly swamped.

Percevale's ship had no time to make the shore,
But grounded itself on the river island
As they were carried inland by the huge tidal wave.

Most the armor-heavy Vikings were drowned
Or swept away, but, towards evening
An exhausted Viking Chief
Began to regain his strength on the island beach
Not far from Percevale and his party.

The tide-wave bore its way past
The Severn Country Inn where Taliesin

Watched and heard the death screams in his mind
While he gazed over the fields of purple heather
Waving in the heartland,
A land now safe for the time being.

It was a steamy metal night on the river island—
Not a breeze was to be felt.
Sweat dripped on one's cheeks
Even as one lay still,
Wrung out by the day's events.

Night fell, and was brightened
By the perpetual twilight that lasted all night
This time of year in Britain and by the bright moon
Which was only a day past full.

"Goodnight, moon," said Percevale.

"Goodnight, Percevale," answered the Viking Chief,
Thoralf, from out of the shadows,
His axe of hot metal already in hand
And gleaming in the moonlight.

"Live and be free," said Percevale, "or die."

Thoralf thought but for a second
And then threw the axe directly at Percevale,
End over end;
It glanced of Percevale's sword
As he knocked it to the ground.
Both men just stood and stared at each other.
Percevale did not yet release his hidden darts.

The Viking Chief gloated,
"One sword is no match for two Viking axes,
But tell me first, before you die,
Just what it was that was a match
For my fleet today. I must know
Before I split your head in two."

Percevale replied calmly,
"Your fleet has been defeated by none other

Than your own abandonment of nature.
Your forgot about the moon because it was cloudy
And since the nights are short this time of year.
There is your conqueror, Viking Chief;
There, in the sky just over your left shoulder—
It hangs there, near Jupiter."

Thoralf chanced a glance at the moon,
Yet, in that instant Percevale retrieved his sword,
And spoke, "Each month the full moon and the sun
Conspire to raise a forty foot tide on the Severn."

The Viking Chief just walked away,
Moonstruck perhaps.
Percevale did not attack him,
But wondered: Perhaps there comes a time
In an evil man's life when he wonders
What he has become
And of the terrible deeds that he has done."

Well, these were the tides of life and death
That swept through men's hearts
From time to time—they could take you
To new highs or lows,
And now and then you stayed there,
And sometimes you fell back.

It was well after midnight now,
And Leo had already duve into the west.
Mars was passing Saturn, and Antares,
The heart of the Scorpion,
Was already rising in the southwest.

{The Ring of Time and Motion}

They soon arrived at the sacred Ring of Stones
That signaled the approach to Salisbury Plain.

Here they rested, just beyond its perimeter,
For the horses would not go near this magic place.
A priestess approached them. "This place,
Stonehenge, was constructed

To the measure and motion of the sun,
The moon, and the stars.

I welcome you, for your hearts are pure and good.
I tend to Britain's calendar and this is a great day,
For the night and day are of equal length,
Thus indicating the start of our new year.
Come join our New Year's feast—
You will witness the equinox upon awakening."

Upon awakening, they enter
The astronomical wonder of stones at Stonehenge
And rededicate their swords to St. Michael,
St. George, to God, to justice, and the British way.

"We'll be back here by the day
That the sun rises directly over the heel-stone,"
Rallies the King, "but in his heart he suspects
That they may never see this place again."

{Farewell Betelgeuse}

Now the night falls on my old friend,
Betelgeuse, as we stand to its deathwatch
To catch yet a few more dwindling light rays
From the long forgotten sunny days.

The big plough, pointing to the universal hub
Now tips and pours its cosmic contents
Down upon us.

Some days, like today, the moon does not rise,
But is up and stays up all night until sunrise.
And if one has good luck, it's cold and clear of sky.

Then find the darkest of night places,
A place that is so dark that you won't even
See the rocks ripping into your new shoes.
So dark that a lantern is needed
To spot a brown rabbit.

And solitude is requisite as well as is the quiet,
So that the river castle trumpets
Can be heard echoing and harmonizing
Up and down the valley.

Then, that's the time to cling to the earth
And look down into the bottomless sky
Of the last winter night
And say farewell to a favorite star
That's setting for the season.

Can you own a star? Why not?
Whether one is young or old,
You can pick out a star to own.

Think of the wealth within a solar system—
It can all be yours—you're rich beyond measure!
And gain a friend at the same time.

Then, on some lonely night,
When you sight your friend of a star,
You will be cheered, for you will know
That you are not alone.

I've tried to own Betelgeuse,
That poor dying red star—
Perhaps long since abandoned
By the people of its planets
As it expanded into their orbits.

Oh, I have tried to own Betelgeuse,
Really I have, but I have found
That it has now come to own me!

She is dying now, my star, having given her all.
Now, she takes—I give!
Never lose you!

Planets already bejewel the upcoming
Summer sky if you stay up late to see them.

Saturn and Jupiter escort Spica across the heavens,

And the great Spring Kite sails high in the sky,
Pulled along by Arcturus, while the Great Hook
Dredges up islands from the sea in the south.
Orion is behind the sun, its last whiff of influence.

The Great Bear had come out from
The winter's lair of the Northern Crown.

{Several Impossible Challenges}

Summer is warm but not yet indolent;
It is the lovely month of June,
Perhaps the greatest month of all
Since many are free now, from school,
And still excited about the upcoming summer.

Victorious knights return from both
The Asia frontier and from the Unknown Sea.
But sadly, St. Patrick has died,
And all Ireland is in mourning.

The KIng receives a summons
From the Avalon Lady of the Lake.

A King may receive a summons
From but one person and place:
The high priestess of Avalon—
The Lady of the Lake herself,
The distant power behind the
Fables and fortunes of Britain—
The Mother Goddess
Who reigns wholly and supreme,
With the assistance of "The Merlyn".

If "The Merlyn" be the power behind the throne,
Then the High Priestess of Avalon
Is the power behind "The Merlyn".

And so our Percevale again takes
The lower shield of the White Horse
And slips unnoticed out of Camelot
During the height of the summer festivities.

He rides to Avalon,
A land forever shrouded in the mist that separates it
From the world of mankind.

Many have been lost trying to cross
Avalon's impenetrable swamps,
So he waits patiently at the edge of the foggy lake.

He brings the Crimson Spear,
For this is surely a gift from Avalon,
As was Arthur's Excalibur
And Price Valiant's Flamberge
(The Singing Sword, or "Flame Cutter".)

Avalon is flooded with water during summer,
And with treacherous ice in winter.
And there is always the fog,
Which only the priestesses can wave aside,
And they, and they alone,
Know the underwater paths for the horses.

Many adventurers have fallen
Into the gloomy depths of despair and death

Trying to find these trails,
So Percevale awaits his guide.

Only once in a great while
Is a King summoned to Avalon,
For most of Avalon's effects are not direct,
But long range, and even so, are often carried out
By "The Merlyn" or "The Taliesin",
The only residents of Avalon who are allowed
To mingle with those of the mortal world.

The guide arrives, and Percevale, without a word,
Steps into her canoe, for she is a novice
And is not allowed to speak.

She waves the mist aside
And they approach a castle in the water
And then enter the Lady's mysterious chamber.

The Lady of the Lake appears,
Old now and perhaps dying.
"Thank you for your rescue of my daughters,
Eve and Melody.
They are my second and third born, respectively,
And may someday have to rule this isle
If for any reason my first-born cannot.
Now, Percevale, name your pleasure
And it shall be yours! Anything you want."

Percevale replies:
"I ask no pleasure but that of continued life.
There is one thing, however—
I should like to gain the power to destroy a witch,
To free those poor souls who are enslaved by her!
For I swore an oath to return there one day
With the power to succeed."

The Lady of the Lake—the Mother Goddess,
Now finally growing old with age
After many centuries,
First speaks to our hero
About age and the ancestry of the Round Table
Before answering him:
"Here in Avalon,
The Royal Line consists solely of women.
Soon, my first-born daughter will take over for me,
As someday, her first-born daughter
Will take over for her.

Only a women can be sure of maternity—
Paternity is never certain;
Who knows who one's father might be!
Thus, a royal line of first-born sons of Kings
Really does not make much sense for us,
But we tolerate it in your world.

Here, we seldom even keep track of paternity,
But, in your case, and in the case of many

Of the knights, an exception was made
In order to try to save the world.

When I was young,
Four hundred years ago,
I played with Merlyn,
Who was also a child at that time.

I happen to know that he and I
Are the great great great grandparents
Of yourself and of many of the knights,
Making most of you third and fourth cousins—
Indeed, Lancelot was born here as my last son,
Thus his full name, Lancelot du Lac.

And so in this way we passed our godliness
On to man in a last desperate hope
Of ending the many centuries of the Dark Age.

"Merlyn" is not really a man's name,
Though it has come to mean his name—
It is actually a title, "The Merlyn",
Of our only male officeholder,
A position that Taliesin will soon inherit and hold.

The office of "The Merlyn" is the only link
Between our two worlds,
Aside from the rare summoning of a King.

I am the power behind "The Merlyn"—
But I cannot interfere in everyday matters,
For then man would not have his freedom, would he?
We can only do long range planning,
Thus your throne and your bleeding spear.
We of Avalon are not actually gods, but Druids,
Descended from the many 'gods' of old.
We are all that is left of the great Atlantis!

Avalon is soon to be forever removed
From the world of mankind—
This we have known and feared—
So, we have passed our legacy of love and goodness
To you and your knightly cousins.

As for your witch, she was once one of of us,
But has since gone astray.
That's how she knew about your bleeding spear
And why she fears you.

As we may not interfere directly,
We may not slay her.
But, you have asked for the power to destroy her
And so we will see that you have it
In the form of your spear
And in the strength of yours and Taliesin's minds.

This but makes you her equal.
Success or failure will still come
From within your own strength and goodness.
But bring the Crimson Spear!

Indeed, you would be doing us and the world
Quite a favor if you were to succeed; but beware,
She once held the high title of Death-Crone,
And she will undoubtedly place a curse upon you.
Just remember this: never give up hope,
And know that every curse has an escape!

But how sad that she was once one of us
And is now out of control!"

{The Curse of the Death-Crone}

As Percevale approaches the witch's land,
He sees the shields and helmets of those
Who came and died before him.

He clutches the Crimson Spear close
And continues his approach.
"Now, Bogar, you wait here
And if I do not come out within two days,
Then come in after me."

Percevale feels the watch of gloom
As he enters the territory of the witch.

Knowing that he is being watched,
He does not turn around to alert the watcher,
But slides quickly and unbeknownst into the woods. Taliesin glides noiselessly,
Silent and invisible in Percevale's mind!

He peers in a window and sees a pitiful sight.
The witch's slaves are from
The world of the deformed and misshapen—
Those who are most easily enslaved.
Plans are made and a good night's sleep is taken.

In the morning a huge menacing giant
Blocks Percevale's path,

But there is something very human and caring,
Yet guarded, seen in the giant's eyes.
Apparently the giant is too large
To fully feel the effects of the witch's drugs,
And so Percevale speaks to the giant softly:
"You could easily escape this spell and be free!"

The giant replies: "You are correct;
I stay only to protect my friends from further harm,
And indeed I will help you kill the witch
If you will but insure the safety of my friends!"

"I am King of Britain
And the safety of all my subjects concerns me.
Just keep your bewitched friends in check
While I do battle with the witch
And soon you shall all be free or I'll die trying."

Now Percevale faces the witch, but not alone,
For Taliesin has joined with him in mind,
And the bleeding spear is at hand.

"'Tis the accursed Crimson Spear from Avalon!"
She cries. "Take it from my sight,
I can not bear to look at it!"

But Percevale holds it all the more firmly

As she tries to wrench it from his grasp
With the powers of her mind.

She fills his minds eye with evil sights of monsters,
But ever still does he hold the red shaft;
It is now bleeding profusely
And its blood is pooling on the ground.

For a day and a night,
The battle of the minds continues,
Percevale and Taliesin barely holding their own
And growing nevermore weary,
Feeling at each instant
That they cannot last another moment.

Meanwhile, no potions are being dispensed
To the enslaved; they drink but the purest of water
And so they are slowly regaining control.

Towards morning, the battle draws to its climax
As Avalon's grandson is assaulted
With every trick known to sorcery
By Avalon's daughter gone astray;
But, Taliesin has studied under the master Merlyn
And Percevale has the strength of ten
Because his heart is pure.

And then it is over.
As the witch crumples to the ground,
Defeated at last.

She finds those last ounces of strength
That come at the time of dying
Using them to place the curse of the Death-Crone:

"Percevale, from death's doorstep,
I, the Death-Crone,
Curse you with my last breath;
I curse you with the worst misfortune
That may befall a man:

That you will never find love
Or be loved ever again—

Until rocks flow like water,

Until the day comes that the sun does not rise,

Until the new moon is seen with the naked eye,

Until the planet Mercury is seen at high noon,

Until fire is seen in water,

Until it snows in Cisalpine Gaul on a summer day,

Until all of the above events happen
On the same day within a month from this very day!

In other words,
You will never ever find love or be loved,
Since, when these events do not happen,
For they cannot happen and be seen by you,
You will not only be unloved nor able to give love,
But you will also find the world to be filled
With hate towards you, and you will soon die
And forever wear the foolscap of eternal shade,
For no man can live for long without love!"

The witch dies, the King is cursed,
But the enslaved are free!

{No Hope for the Hopeless}

Bogar, forever dedicated,
Takes what is left of his master back to Camelot.
Bogar notes the King's despair
And so Percevale tells him the tale of the curse.

"I shall never succeed, Bogar,
For most of the witch's challenges are impossible;
That's the joke of it, I guess.
She just threw in one easy one,

'When rocks flow like water' to give me false hope,
For I do know of a place
Where rocks flow like water.

But no one has ever seen the new moon.
Of course, the full moon is easily seen because
It is completely lit on the side facing us
And rises when the sun sets and is therefore
Up all night, but the new moon is just the opposite:
It rises in the morning, is up all day,
Sets at evening,
And is lit only on the side away from us.
It has never been seen, Bogar!

Oh, we have seen the slivers
Of the very young and the very old moons,
But the new moon gives no light at all,
So, even if we see but a thin crescent moon,
Then by definition, it is not the new moon.
Even if we knew where to look for it in the sky, Which we
don't, there would be the glare of the sun.

Even the stars, which do give off light,
Cannot be seen in the daytime,
Even in areas of the sky not close to the sun.

And Mercury, being so close to the sun,
Can only be seen just before sunrise
Or just after sunset, but never at high noon!

As for snow in late June or July in Southern Gaul,
It is not likely and has never occurred.

And I have not yet known a day
When the sun did not rise.
Even on cloudy days
We know that the sun has risen,
For there is light behind the clouds.

And fire in water! It cannot be.
Water conquers fire, they cannot coexist.

For any of the above to happen is impossible.
For all of them to happen on the same day
Within a month is beyond impossible,
Yet, I will not give up hope for I know
From Avalon that all curses have an escape."

Percevale spends the day
In the archives of Camelot with Taliesin.
Then they spend all night
In the Merlyn Tower Room,
Where they pore over old manuscripts
Full of diagrams,
But only this much becomes known:
The new moon is to appear in two weeks—
This fixes the day;
And there is only one place
Where the rocks flow like water—
This fixes the place!

There is hardly time to get there,
So the King immediately leaves for Iceland.

{The Ice Maiden}

The chronicles covering the first week of
The journey have not survived the ravages of time,
So we find ourselves already close to Iceland.

The sea is glorious and the air is fresh and pure.
We do know that during the journey north,
The twilight lasted longer and longer each day.

There is not a moment to waste, However,
Percevale spots a vessel in distress behind him,
And for just a moment he wonders
If he should take the time to come to its aid.

But, there is no real choice, so he turns back
And although her ship goes under,
He manages to pull her from the depths
And spends over an hour reviving her.

And, even when revived,
Her lips will not part from his,
For they have tasted each other
And found it to be sweet.

"I am cursed, you cannot love me,"
Says the Ice Maiden finally, named Dheryle.
"I am sent to remind you
Of that which is forbidden to you!
I have no choice; the spell overwhelms!
You should have let me drown;
Then you might have had some peace from success.
From now on, everyone you touch
Will catch the curse until
The world fills with hate and destroys itself."

"So this is how it is going to be,"
laments Percevale.
"How I shall hate to give up life's wonders
When I am gone!"

{The Greatest Day on Earth}

But, this is to be the day of the new moon;
At least there is a chance, thinks Percevale.
They arrive on the shore of Iceland,
And, on this day, as on every day for a month
Either way in this northern land,
The sun does not rise,
For it did not set the day before.

Just before noon, strange bands of shadows
Begin to rapidly cross the land
And Percevale feels that perhaps the end is near. The
ground begins to shake and heave
For a few moments and then all is silent,
So very silent as to strike one dumb.

Something terrible seems to be happening.
Grazing animals look for shade trees
And lie down to sleep.

Then, about noontime,
The shadow of darkest night covers the land
As the moon begins to kiss the sun and cover it—
It is a solar eclipse!

Merlyn's old notes in the archive were accurate!
Thank the gods for the old wizard!

During the seven minutes of total darkness,
Percevale sees a black disk in the sky,
Surrounded by faint wisps of flame.

It is, of course the new moon in all her black glory;
Indeed, the new moon can only be seen
During a solar eclipse, and never at any other time.

And there near the sun is a bright 'star'.
It can only be the planet Mercury!
Yes, there it is, in plain sight, at high noon.
And farther out, Venus can be seen!

Now the ground begins to really shake, and
Percevale hurries to his ship with the Ice Maiden.

They leave Iceland but see the volcano erupt;
Rocks are flowing to the sea like water!
But, the water puts out the fiery flow
And so they do not see fire in water,
But just a lot of steam.

Then a tremendous plume
Of smoke and debris is sent up into the sky
And is carried south by the unusual winds
Born of the marriage of summer warmth
And ice cold air brought on by the blockage
Of the sun's rays by the dense volcanic ash.

The spontaneous cold front sweeps south to Gaul
On the reversed upper winds,
Bringing the darkness of the ashen sky with it.
As no sunlight can penetrate,
The air below grows colder and colder,

And what would have been rain now turns to snow
Over Cisalpine Gaul for a brief time
Before the westerly winds can disperse
The volcanic cloud around the earth.

That evening the sun sinks low, but does not set. On the
water is the glitter path of that fiery ball—
And so we have fire in water!

The sun has kissed the moon,
And Percevale gathers the Ice Maiden
Into his arms and kisses her,
His capacity for love far from dead,
But growing stronger every minute
Of this glorious day
As both of their curses fall by the wayside.

(Taken from the Celtic Chronicles,
Found in an iron box beneath an Abbey.)

{The Last Curse on Earth}

Percevale returns home and sits down
To hear the Giant's tale and the giant begins:

"The witch placed a curse on me as well.
I will forever roam the earth in sadness
If I do not accomplish the following
By the end of this day:

I must see the sun set three times in one day,
And, I must, during daylight,
Create a dark space behind me that never ends.

What will I do? I cannot stop the sun
And raise it up again,
Nor can I cause the absence of light behind me
And into the infinite depths of space!"

Day is nearly done and so the horizon
Is rising to meet the bloodshot eye of day.

Percevale quickly leads the giant to the shore
Where a small piece of low hilly land
Juts out into the sea.
They face to the west and view the setting sun,
Now but a symbol of the sad giant's dying hopes.

The sun drops though some clouds
And is bright again,
But half of it is already below the horizon!

"Look at your shadow, giant!
How long is your shadow at sunset or sunrise?
What is shortest at noon grows longer
As the afternoon wears on, until finally,
It stretches forever behind you,
Since you are directly between the sun
And that which is behind you."

"That is fine Percevale, but the sun is nearly set
And will certainly not rise again until the morrow.
I must still see three sunsets!"

"No time to explain now, giant.
Quick! Lie down on the ground
And see your first sunset today
As the top sliver of the sun falls below,
And is extinguished by, the horizon.
See! There it goes.

Now, quickly, stand up to your great height
And what do you see?"

"I see the tip of the sun again!"

"And your second sunset of the day, giant?"

"Yes! I see it, and another green flash as well!"
"Now run up yonder hill and bring up the sun again
So that it may set three times in a day!"

The gleeful giant runs up the hill in great leaps
And turns to see the sun set three, four,

Even five more times,
Each sunset lasting but a few seconds.

< 37 >
— The Root of Ill and Evil —

Low serotonin stems from genetics,
Stress, lack of exercise, or the wrong foods,
And can cause anger, anxiety, and
Depression, even bad behaviors, crimes.

< 38 >
— Chemical Imbalances —

Since the aggressive urges leading to 'sins'
Are not caused by (D)evil, the 'sinners' are not
To blame, although we still have to lock up
The violent ones to protect ourselves.

< 39 >
— A Primary Stressor —

Prolonged coercive job stress can damage
The immune system and brain transmitters,
And cause secondary time-stresses,
Sleep deprivation, and social problems.

< 40 >
— What Sleep is For —

Sleep deprivation can cause accidents,
Immune system damage, and subsequent
Poor physical/mental health, as well as
Memory loss—so, get your eight hours in!

THE GREENLESS WORLD

I'd come to this strange and foreign world
Over three years ago as a scout
For a phosphorus mining expedition,
And here I had remained, marooned,
For the nearest asteroid supply bases
Had been closed for lack of
Their necessary Earth supplied material.
Well, at least I had life. I'll take that anytime.

I was thankful, too, that my alien friend,
a native of this planet, was female,
And that we were compatible
Both genetically and physically,
Although we were probably unable
To produce offspring—at least so far.

Science long ago had indicated that the Earth
Was probably not the birthplace of mankind,
That Earth was seeded by ancestors
Who were common to all the galaxy.

My friend's name was Serena,
That being the closest English translation.
Over the years here,
I had learned her language
And she had learned mine.

We lived together 24-7,
And so I had been spared
An eternity of loneliness,
Although it had been a very close thing,
I being the only human here
And she being one of the few remaining natives
Of this doomed planetoid.

This planetoid had been dying since its birth,
For it had three suns, one of which was always shining,
And so it was only a matter of time, I suppose,
Before all the underground springs evaporated.

But, these types of planetary events
Were still measured in centuries, if not more,
And there was perhaps
No immediate danger in our lifetime,
Although life here was certainly
Becoming more difficult;
Hence the already great exodus long ago
Of those who could afford to leave.

There was never any darkness in this land
Where the sun always shone,
Not even inside the caves,
For the phosphorous in the walls and the ground
Gave off a constant luminosity.

This phosphorescent light
Had been hard to get used to, at first,
Although Serena had no problem,
Having been born here—
She even had the natural ability
To sleep with her eyes open.

It was the hottest part of the week now,
The time when the two largest
Of the three suns shone at once,
There often being such an overlap
As this for days at a time,
And so we often had to retreat
to the "cool dampness" of our cave—
Our home in this primitive world.

And even when there was but one sun in the sky,
It was still quite unpleasant to be outside,
For it was always hot, and bright, too,
For the suns, all of them, were large,
And one could not easily look up
Into the sky near any one of them.

As I said, the cave was lit
By the radiant glow of the walls.
No real blackness anywhere.

Our lunch was boiled brown vegetation,
The only cuisine available,
However, when one is hungry,
One is thankful for anything at all.
No gourmet food here.

Serena had never known darkness, and, indeed,
There wasn't even a word in her language
Which meant anything close to "dark", "night",
Or even "black"; however,
I'd been able to convey the concept
By using the absence of light as an analogy.
Of course she still had trouble grasping the idea
Of "that which could never be".

I suppose it was like asking someone
To visualize a color that one had never seen.

Naturally, I tried covering her eyes
To simulate darkness
(Since she couldn't close them),
But, she still reported a yellowish color,
And later, upon inspecting her eyes,
I noted that they gave off a cat's eye type of glow—
Just like every damn phosphorous rock on this planet.

Even the sand shone like gleaming yellow snowflakes.
Ironically, this was what had brought me
Scouting here in the first place—
The prospect of mining that rare yellow light
That made fireflies glow
And caused those struck matches of old to light up,
For the Earth's supplies had long since run out.

This ever present light was, at first,
Psychologically disturbing,
But, I'd learned to live with it,
First by sleeping with a band of cloth
Wrapped around my eyes, although, gradually,
I lost all track of time and just slept
Whenever I got tired.

I also had to be careful not
To come into any rough contact
With the phosphorous rocks in the cave walls,
Lest they should burst into flame.

Yes, it was a rather precarious existence,
Though a liveable one, but, alas,
I could never go home again,
For the Earth had been destroyed
By a giant comet,
One of the Perseids, the shower
Whose many precursions
Had given us the wonderful
Meteor shows of that name.

I turned to Serena and spoke to her about it,
Having been unable to deal with it until now,
And because she had only recently gained
The scientific knowledge to be able
To understand solar system concepts.

"I was one of the lucky ones, Serena,
For I was already out in space,
Had just recently launched, in fact,
When the disaster hit my home planet, Earth.

"You cannot know the shudder that went through me
When I realized that all that I had loved was gone,
That all that it was or could be, all that had formed me,
Given rise to me, was gone forever.

"My Earth was one of the most priceless work of art
That the universe had ever known.
This rock on which we now live
Is not even Earth's pale shadow—
At least we do have pale shadows on your planet,
though they are hardly noticeable.

"At first, when I saw Earth's fireball,
I thought that I had seen a shooting star, but then,
Noting the origin and size of the spectacular explosion,

I was overcome by a horrid feeling—
One that was chill and sickly like any I'd never known—
that it was indeed the Earth that had left us.

"I could do nothing but continue on,
For the Earth had no equal in our solar system.
Oh, we had long ago searched
The entire galaxy in vain for such a paradise,
But the Earth had remained unmatched."

Serena thought for awhile,
Having only recently grasped the idea
Of a universe filled with worlds,
She never even having seen stars
In this land in which night had never fallen.

But, again I was fortunate that
She had an open and intellectual mind;
So, during our recent studies
I had been able to take her thoughts and her mind
Across many centuries of learning and knowledge,
Sequentially educating her, step by step,
Using small and primitive learning blocks
Until reaching some rather complex theories.

She was now able to understand such concepts
As solar systems, space travel, physics, biology,
And many other unseen wonders
Like oceans, rivers and lakes,
Which, though quite impossible on her planetoid,
Were at least conceivable to her,
Since she had often seen water bubbling up
From the hot springs—which, by the way,
Were apparently the limited
And unrenewable source
Of both water and oxygen
On this planet.

Finally, she spoke,
After allowing the cloud of sadness
To pass from my brow,
For she was emotionally very capable,

"Patrick, you lost everything that day—
You are a man without a world.
How many people died? How many survived?"

"Trillions died—that is a number you don't have here,
But take your "deca" and multiply it by itself
For "deca" number of times
And you will be close to knowing what a trillion is.
No one on the ground survived."

"A trillion is like the number of grains of sand
In the desert outside our door," she answered.

"Yes, Serena! I might of said that in the first place
But I suppose I've been too much of a scientist lately.
As for how many survived,
I'm sure that's only in the thousands—
Perhaps eventually only in the hundreds,
Since many Earth outposts were contained
Within domes on uninhabitable moons and asteroids
And were quite dependent on the Earth
In the long run for their survival."

"I understand more and more everyday,"
She answered,
For she was now quite proud
And even happy with all the ongoing revelations.

"When you first fell from the sky
I thought you a god,
But now that you explain everything
I see that it all makes sense,
And what once seemed magical and clever to me
Is now all laid bare before my eyes
As something entirely reasonable."

She spoke mostly in English now,
There not being enough words
In her native tongue to suffice,
But, of course,
When discussing particulars
Known only to her world

We had to use her language,
Which, for example,
Had hundreds of different words
For all the various kinds of light and heat,
Although none for weather,
Since it never rained or even got cold;
And, as I have mentioned,
There were certainly no words for night,
Blackness, stars, or for other worlds.

She continued,
“We have been together several years now,
And still I awake each morning eager to learn
Of new mysteries. Is there no end to knowledge?”

“Oh,” I replied,
“Where I come from there is truly no end,
But one cannot possibly know everything,
So one ends up finding out things
Only as they are required.
Oh, the wonders I could have showed you on Earth:
The colors, the mountains, the forest, the meadows,
The scents, the tastes, the inventions.
I’m sorry that I don’t have any books with me
Or even something so amazing as a mirror to show you.”

“A mirror?”

“Yes, you can see yourself in it.”

“See myself? See another me? I cannot.”

“Yes, it’s like a reflection in the water—
Oh, I forgot—there is no standing water here,
And damn, I don’t even have a shiny belt buckle
To use to show you the effect,
And all the glass in my spaceship is non reflective.
Anyway, yes, you could see yourself
Just as others see you.”

“From the outside of me? I sort of understand
But I cannot quite imagine.”

"When mirrors first appeared on Earth
In the form of polished metal,
People thought them magical,
And even in modern times
One could watch with wonder
The amazement of babies or small kittens,
Who, though they both quickly got used to it,
Thought first that they'd seen
Another of their species."

"Kittens? Cat?"

"Small furry animals.
Domesticated—meaning tame or not wild."

"Animals? Wild?"

"They are other forms of life,
Some with four legs,"
I explained, ever so patiently,
For there were no animals on her planet.

It was in this way
That we often got nowhere fast with words,
But then, all of a sudden progressed
With great leaps and bounds,
Especially with material ideas;
However, abstract concepts took longer,
And concepts like darkness
Were still pretty much incomprehensible to her.

"We had animals in the old days," she said;
"There are drawings on some of the cave walls
Of such as you speak.
They are all gone now, like your Earth.
You seem so sad when you speak of Earth.
It must have been wonderful.
What do you miss the most?"

I thought for awhile,
Thinking of the scorched surface
Outside our cave.

"What I miss the most is not the darkness,
For I can simulate that here when I sleep,
And not love, for I surely have that now with you,
And not the cold, for I never liked it,
Nor life, for I am happy to have it here,
If nothing more;
But, what I miss most,
If I had to say some particular thing,
Is the color green,
For green is a color
Which does not seem to exist here,
The hue that is the soothing and lush
Life-giving restful green of Earth.
It was said the be the sanest color,
Evoking serenity, as in your name."

"What is the color green?" she inquired.
"I know the blue sky, the golden suns,
The tan rocks, the brown leaves
And the brown vegetation,
The pink of your hidden parts,
The red of our blood and of your hair,
the orange flashes of fire,
The darker brown of trees that almost suggests
The strange black color that you speak of,
The gray shadows, and the yellow of phosphorous,
But I have never known there
To be a color called green.
What is green?"

"I wish I could show you, Serena,
But there is no green on your planet,
Not even a tint or a shade of it.
On Earth,
The leaves and the vegetation are mostly green,
But here, the same are all born brown,
Even in the shadows of caves.

"Some people on Earth have green eyes even,
But, alas, mine are brown,
And there is no other body part which is green.
Although nothing much else on Earth is green

But the vegetation, green has,
Even more than the blue of sky and ocean,
Come to be regarded as the sweetest color on Earth,
For it represents all that is living and supportive of life.
It is very calming and serene, like you,
And therefore many people use it
As the color of their carpeting.

"Many of the other colors have drawbacks
Or more specific uses:
Red, for example means danger, blood,
But having red tablecloths in eating places
Makes people hungrier and so they order more food;
Pink is debilitating, and so many of
The game playing sports teams have painted
the visitor's locker rooms in that hue;

"Blue is energizing and is
Often used in working places;
Yellow is bright and cheerful, the sun's color,
And is often used in cooking rooms called kitchens,
Although yellow can also mean caution, danger
Even, especially with black,
As on stinging insects called bees;

"Purple is used for mourning death
Or for the regal Kings and Queens, the rulers;
Our brown, like all around here,
Is actually the most popular non primary color
And is not, therefore, even in the spectrum,
For it is made up of red and yellow (orange) and black."

"But," she persisted, "what is green like?
If you can't tell me what it is,
Then maybe you can say what it is like,
Or perhaps you can say what it is not like."

"Either way that's hard to say,
For green is a unitary hue, and also primary,
And, so being, means that there is nothing like it,
No overlap; although if I had to say so,
I think green is more like blue than any other color,

But I only say that because green
Is a cool and soothing color like blue,
And not a fiery color like red, orange, or yellow.
But, I should tell you that blue is certainly not green,
Nor is green blue. If I myself had not known green
Then I doubt that I could have
Conceived of its existence."

"That is fine philosophy," she said,
"But it does not tell me what the the color green.
Have you any green clothes?"

"I do, or I did, but they're not with me—
Not even the slightest thread,
for I've already examined
All my clothes and space suits.
It's not that I don't like to wear the color green,
Although it is seldom worn on Earth,
Except on St. Patrick's Day—
I guess there was already so much green on Earth
That we all came to prefer more of a contrast.
And my spaceship, it's all metallic skin
And fiber optic conduit—
There's no green anywhere in it or on it."

"Patrick, I really wish that I could know green."

"Too bad there are no rainbows here or waterfalls."

"Rainbows?"

"Caused by water vapor falling from the sky,
Called rain, or fog or moisture
That divides white light into the colors
That join to form it."

"Waterfalls? Falling water?"

"There are none to be found here—
The rivers dried up long ago,
Leaving only those ruts in the ground
That I sometimes call canals.

I really miss green, though—
It is somehow a part of me,
But let's not give up on it so easily, Serena."

We ventured outside the cave
To begin a search for green,
There being only one sun up now,
Although it was still a scorching 130 degrees outside.
We lifted some rocks,
Finding various insects thereunder,
But they were either brown, gray, drab, or colorless.
I next peeled back some bark from a 'tree',
Hoping for some inner tint of mossy green,
But it was only tan.

I pulled up a plant and tore open a leaf,
But had no luck with that either—
Evidently chlorophyll played no role on this planet.
Next we cracked open a rock
But found not the fabled color.
This planet was indeed a greenless world.

"I guess green is not necessary for life here," she said.

"I guess not.
This reminds me of the time I read a book
Which did not use the letter "e",
The most frequent letter in our alphabet.
I didn't even notice it at first,
Although I had a vague feeling
That something wasn't quite right."

Even though I was now very tan,
I didn't dare stay out in the sun too long,
So we gave up our green search for today
And headed back to the cave.

"I wish I had some ice," I said.

"Ice?"

"Solid water—hard as a rock,

Nut so refreshingly cold."

"Make me some ice, Patrick."

"That I can never do in this climate.

It never gets cold enough."

"What is cold?"

"It's hard to explain if you've never felt it,
But it's sort of like how the underside of a rock
Does not burn your hand because it's cooler than the top;
Only "cold" means much much more cooler."

"I know not this thing called cold," she replied.

I looked fondly at Serena
As we walked back to our eternal cave.
Her skin was a very deep bronze, all over,
For she always went naked.
She had blonde hair,
But no other body hair
(since her race had never known cold, I guess).
Her hands and feet were wide and leathery,
With six toes and six long slender fingers.

She had no real eye lids to speak of, or eyebrows,
But otherwise had all the other humanoid features
And anatomy—she was a human first cousin, perhaps.
She was completely vegetarian by necessity.
She had not a violent thought in her head,
Having, I suppose, no natural enemies to fear.

For dinner we gathered some brown leaves,
Which evidently contained
All the nutrients that we needed,
Since we were still healthy,
And then retired for the 'day'.

"What else do you miss from your Earth?"
she asked as we lay together.

"Well, at night, for this is night to me now,
I miss the stars—those suns of other worlds
That I told you about, the stars that shine
Across the blackness of space,
For they are far away and thus appear very small."

"You mean that the space between the stars is black?
At least I think I know what black might be.
And the suns are not bright and blinding
Since they are so far away?"

"Yes, but we will never see the stars here—
And how I miss them all,
That night sky full of lights.
I used to look up at it as a boy
And dream of going to the stars.
Then the transwarp drive was invented
And my dreams took wing."

"What do stars look like exactly, at night?"

"They are very small, just points of light really,
But they twinkle like jewels,
Such as diamonds and sapphires,
And some stars are even emerald green—
Like the close star companion of Sirius!"

"Jewels? Diamonds? Emerald? GREEN!"

"Jewels are stones that give off light
In colorful gleams and sparkles,
Like when you cover your eyes
After seeing a bright light,
Or like the gleaming sands outside."

"I see them in my mind.
Are there a lot of stars?"

"Trillions—so many that some areas
Of the night sky appear as cloudy white patches."

"I wish that I could see stars, Patrick."

"There are so many things that you'll never see!
If only I'd brought some photos of Earth with me!"

"Photos?"

"Yes, they are like permanent mirrors."

"I could see Earth and yet not be there?"

"Yes, and it would look about the same.
We even have three-dimensional holographic pictures."

"Oh, that there are such wonders!"

As I fell asleep in her embrace,
I had some dreams of Earth—
At least I could still go there in my sleep.
Oh, how often had I taken Earth for granted,
Not appreciated it when it was there;
I even left it time after time
To go off into the cold and colorless void.
I tried to forget it, but I could not.
Well, at least I was alive in this strange sort of Eden.
Anyway, life was life, and more and more
I realized that I didn't really need
Anything fancy from life,
Except love, of course.
Yes, love was enough—
And it is reason for all that we do.

Well, my spacecraft was still in working order,
But was there any place out there left for me to go?
Was there any life still out there?
Or had it all withered away by now
For lack of support?
There were plenty of
Borderline class-K planetoids around
Like this one, but, unfortunately,
None of them were anything like Earth
For a long ways in every direction.
Perhaps I could head in some chance direction,
Running out of fuel, of course,

But coast at a high speed for years—
No, it was too risky—
My ship was only of the intersolar type
And was not meant for distant star travel.

I could, though, return to the main mining base,
An artificial world built on an asteroid;
But, no good,
For it too was a world with an uncertain future,
A world even more sterile than Serena's planet.
No, my life was here now.
Anyway, all the greener worlds
Had nasty diseases and organisms
Against which I could never be immunized.

Yet another of those endless
Overly tropical days dawned,
But only in my mind,
For the sun never rose or set
Without another sun
Already in the sky before it.
No dawn, no dusk, no half light.
I had brown leaves for breakfast again!
Talk about the simple life!
I was becoming a modern day Thoreau.

"Tell me again about the mysterious colors
Of black and green," she asked.

"Black is easier to tell of,
So let's start with it,"
I answered.
"Black is the absence of all color
And so it is the opposite of white,
Which, amazingly, is the sum of all colors,
Although white reveals not a one of them,
Wxcept through a prism.
Since your life is based on phosphorous,
You see a dim yellow even as a background color
When you shield your eyes.
But when I close my eyes
I see a black background."

"I can't think what would be there
If not for the yellow glow."

"I'm afraid I'm not doing a good job of explaining."

"Try harder," she encouraged,

Then it hit me! "What a fool I've been," I said.
"I have an old-fashioned ink pen
Somewhere in my spaceship,
One that writes in black
On something old called paper!"
At once I retrieved it,
And the pen still worked
As I ran it along my skin.
"That is the color of night, my dear.
This is black. See how dark it is?"

"I see now," she said. "It is as I thought,
Being the limit reached by the removal of all light,
The color hinted at by shade and shadow,
The color just past brown, at least for me—
A lack of color really, like you said;
But I still do not understand what is green.
Do you have any green ink?"

"No, people don't usually
write with green ink.
Red ink, yes."

I quickly ran outside
And went through my spaceship
With a fine-toothed comb.
The ship was all white and metal gray;
There was not a shade of green to be found
Anywhere on it or in it.
The seats were simulated leather
And all the electronic readouts were orange on blue.
All the supply kits were yellow
With the insignia of the mining company.
I had trouble even finding anything blue on the ship.

"We're a long way from Ireland,"
I said, exasperated.

"Ireland?"

"It's a country on Earth
Known for its forty shades of green."

"I wish that I could see Ireland, Patrick."

"None will ever see it again except in memory,
Although I came from there."

"Can you not make me the color green somehow?"

The question struck me dumb,
For it was really a very good question as asked.

"Wait a minute," I said.
Green is made from mixing blue and yellow—
But, unfortunately I don't have any paints
Or such mixing materials,
Although I do have a bluish-black pen,
But of course not a yellow one,
For who would write with yellow ink."

"Worse than writing with green,"
She added, smiling, catching on.

"Yes, writing with yellow ink is silly
But yet, there must be some way to produce green.
Serena, can you logically mix blue and yellow
In your mind and then imagine the result?
No, forget it, that doesn't make much sense,
For yellow and blue give no hint
Of the resultant green like, say,
The way yellow and red readily hint
Of the resultant orange."

We sat silently for while, stumped,
The heat growing stronger all the while.

She looked down at the ground, disappointed.
I, too, looked glum for a time. Neither of us spoke.
We saw nothing but yellow phosphorous
And the yellow sand gleaming even more golden
In the light of the twin suns—
There was yellow everywhere we looked—
Hot warm glowing yellow and more yellow
Until it had fully saturated the eyes and the mind.

Then I noted a flash of inspiration on her face.
She smiled and suddenly looked straight up
Into the bright blue sky, as none had ever dared to,
Then covered her eyes and screamed with delight.

"I see green, Patrick!", she cried.
"I see it. It's green!"

I quickly did the same as she, and, yes,
The mixing of the blue sky
With the yellow afterimage of the phosphorous ground
Had produced a clear and vivid green.

She had, at last, seen the verdant color of Earth.

< 41 >
— The Unapprehended Source —

Hostility can stem from low serotonin,
When we fall for our thoughts, hook, link, and sinker—
Kicking the cat or the kids when we get home,
Rationalizing that they made us annoyed.

< 42 >
— Obsessions —

Cravings are not free will, as much as one
Might like to think so. These imbalances
Cause much senseless irritability,
Nail biting, or the urge to overeat.

LIFE AND CONSCIOUSNESS

These are the two
Really last frontiers of knowledge.

We have them surrounded, though,
Since substances go in
And life and consciousness come out.

Life is relatively "easy" to figure out,
Compared to consciousness,
Since it's composed of functions and processes
That we can even see, given enough research;
So, the elan vital has gone the way of the dodo bird
And doesn't compare at all
To the mysterious quality of consciousness,
Which I'm including here since it allows life
To observe its comings, going, and doings.

There's no doubt that life and consciousness
Require a certain kind of physical process,
For life correlates to the physical functions,
As does the contents of consciousness correlate
To cognitive processing.
Note that I said the "contents".

As life is "easy" to explain,
This leaves but the quality of consciousness
To be nonreductive to the necessary physical;
So, then it must be, for now anyway, that:

In certain kinds of biological systems,
The information from cognitive processes
Has a dual nature, like "It from bit",
This being called naturalistic dualism,
A nature that is just as funda-mental
As mass-energy, and space-time, etc.,
It always leading to a mental representation
Of the information in the brain beneath it.

THE VACUUM

I'm really not following this theory.
Remember it has to be simple
AND understandable.
So what makes the vacuum
In your theory to fluctuate?

And if there is something in that vacuum
To cause it to fluctuate what is it?
You just can't say a nothing vacuum
Fluctuates all by itself.
There is a principle out there
Called cause and effect.

It's more like that nonexistence can't be,
As we've always shown; so, something must,
And these are the vacuum fluctuations.

Perhaps we could consider them
Winking in and out of existence,
But the virtual particles can become
Rather enduringly real (or fall back in).

Note that the bottommost fluctuations
Could not not have cause,
Although I have kind of given one;
So, then, the really bottommost causeless "law"
Is then beneath the fluctuations—
You could call it
"Something had to be, since nothing can't"
As being the Why,
With the fluctuations being the How.

Then the What and the Where becomes
As a space-time full of moving stuff,
Of which movement Time is born,
And that's what grants the When, Now, and Then
Of future, present, and past.

All of this combined
Is a so-called "spirit of life"

That can form the Who,
Which is being.

What may be hard to understand
Is that causes can't go on forever beneath;
Yet, how could they, for I've never met
An infinity
Or an eternity
Which could actually complete itself?

Vu = Uniform velocity = Higher symmetry...
Vr = Random velocity = Asymmetry...
Va = Absolute velocity = Highest pure symmetry state...

Austin, extend finite decay to its absolute limit,
At near infinity—Think about this—
All particle structure of Vu decays to Vr of the Va...
At Vr of Va symmetry of frequencies must form
From such decay at Va limit.

Va at limit's super-symmetry—
As it exceeds its Vr—
Must form the highest/ lowest thermal/ entropy
Quantum state of hydrodynamic
(The highest compression state
Of infinitesimal field matter-higgs field mechanism) =
The prime mover,
To recycle the finite decayed Vu beyond Vr,
Back into Vr from the Va's Vu,
On into a new structured matter of Vu.

This is infinity and eternity completing itself—
All by its lonesome,
By the third stage of necessary
A priori modal logic's universal self-mechanics—
It's a self-re-boot-strapping Universe...
All this is the cycle of the dynamics
Of FS (Fundamental Substance)
It is because it simply has to be—
Tain't no other logic possible...

Yes, that would work fine
To keep our universe going forever,
Ending and restarting, if it doesn't thin out
And I wish to leave all of that intact
Since it is completely workable.

I've introduced a pre-stage,
That of the fluctuations
In the larger expanse of the Cosmos.

Think of another universe starting up.
The fluctuations produce
The fundamental substance(s)
In a positive/negative balance
Through the emission of particles pairs,
Some of which stay around
And then do as said
To perhaps have a cyclic universe.

One reason for this pre-stage
Is that it gets us around the paradox
Of why a specific amount of fundamental substance
Would have been sitting around forever
With nothing prior to account
For that certain amount.

The second and third reasons,
Ones that match the thought experiment above,
Are that there do seem to be vacuum fluctuations
And that the positive energy of the universe
Seems to balance out the negative.

We also get around there having to be
Any special time or place for a universe of stuff
Without anyone deciding where or when,
Since one could be anywhere, any time.

Same for anything else
That would have to be not special,
Excusing very tiny sizes, maybe,
If it is that the smallest workable size
Is an absolute measure because it has a limit.

So, if we join the pre-stage to the FS
Then we have it all.

So, then, as in our cyclical universe's instance,
As well as in any other universes
That were able to go somewhere
On up into complexities
Without some kind of inertness or a flop,
Just about all that's possible can happen,
Such as our human life or similar, better, worse,
Or just different, for all of forever remains.
And, too, all of this has been going on "forever".

It all shows that there's nothing outside
Of the causeless causings of all that becomes.
Existence <u>must</u> be so, which is not surprising
(As if Nothing could produce anything).
Nor any "God" for that Guy would certainly need cause
And could not be more Fundamental than his parts.

On average, the universe sums to nothing,
As the negative potential energy of gravity
Cancels the positive kinetic energy of mass-energy,
As well as electrons canceling positrons, etc.,
But this is not the end of the accounting,
For their was/is the capability of the fluctuations
That emit the opposite particle pairs—
And this is a something and not a nothing.

As for actual real substance
Always having been around,
We can't have a causeless eternal substance
Of a certain amount around forever
With nothing deciding the amount.

This eternal substance idea, though,
Is one of my other TOEs,
As I have three of them.

The third TOE is that everything happened
Within the the wave function of the entire universe,
And still may be, but I'm not so sure about this one.

Even if we retreat to that "we can't know"
The results are still the same that we are free to be.

"God" is not a simple answer,
For S/He would be the most infinitely
Complex composite system of mind
That one could ever come up with.

1. Possibility

Particles and Forces.

2. Probability

What best combines and continues.

3. Necessity

A total lack of anything is impossible.

< 43 >
— The Feedback Loop —

Persistence of thought, known as obsession,
Is enabled by low serotonin:
That same negative thought breeds depression,
That same anxious thought induces panic.

< 44 >
— The Same Root Cause —

Odd that Paroxetine can cure compulsion,
Panic, anxiety, and depression?
No, for obsession is the casual root—
As all involve the persistence of thought.

< 45 >
— Stick-to-it-ive-ness —

Persistence of thought is the curse that leads
To great accomplishments through obsession—
The driving force behind the creative arts,
That driven sense which cannot be denied.

< 46 >
— Just a Short Term Fix —

Alcohol can raise serotonin in
The short run, but decreases it long term—
A doomed attempt at self-medication,
But hold on—relief is being researched.

< 47 >
— Digest This —

Born without the enzymes that digest milk,
She supplies it via Lactaid tablets.
Born without the zymes that digest mood,
Zoloft prolongs her mood regulators.

< 48 >
— The Long Way —

Behavior modification can raise
Serotonin, too, though not as quickly
As medications, like Paroxetine,
But it still works—it just takes longer.

< 49 >
— Learning is Brain Re-Wiring —

Behavior modification can change us, too,
For the worse, if we see too much aggression,
Or do the same thing too long, such as overtime—
Getting well grooved into that same old rut.

A BRIEF HISTORY OF ALL OF HISTORY

The Cosmos

For all time
Vacuum fluctuations
waver in and out of existence
since nonexistence cannot be.

Our Universe

1E-43 seconds
Planck era.

Cyclical compactfication
or a
Vacuum fluctuation eruption

1E-36 seconds
GUT transition.

Strong force separates
from the Electroweak force.

1E-36 seconds
Inflation begins.

Slow rolling scalar field
generates negative pressure
causing exponential expansion
of space time.

Doubling time: 1E-36 seconds.

Vacuum energy density: 1E73 tons/cm^3.

Quantum fluctuations lock in
nearly scale invariant 1E-5
variation in energy density.

1E-34 seconds
Inflation ends.

Decay of scalar inflaton field
causing reheating.

Is this the let there be light moment?
No, photons don't exist yet,
but other massless vector quanta
like left and right weak
and B-L particle may exist.

Things are not well known about this era.

1E-34 to 1E -8 seconds
Quark era.

Quark gluon plasma.

Quarks and super particles
dominant matter content.

1E-17 to 1E-15 seconds
SUSY breaking.

Super partners acquire mass
with the LSP expected to have
a mass of about 10 Tev.

(In induced Gravity model,
this is where mass energy
first generates the
induced gravity field,
Gravity is born.)

1E-10 seconds
Electroweak transition.

The Electroweak force,
under the action of the Higgs mechanism
breaks symmetry.

The photon is born.
Standard model particles get mass.

1E-5 seconds
Quark confinement.

The QCD vacuum becomes superconducting
to color magnetic current.

Quarks and Gluons are confined.

1E -5 to 1 E-4 seconds
Hadron era.

Hadrons are formed:
protons, neutrons, pions etc.

1E -4 seconds
Hadron annihilation.

A brief period of proton/anti proton
and neutron/anti neutron annihilation.

A slight favoring of matter over anti matter,
possibly locked in by CP violation
at Reheating causes some protons
and neutrons to survive.

1E-4 to 10 seconds
Lepton era.

Following Hadron annihilation
Leptons are the dominant energy density.

1 second
Neutrino decoupling.

Mass energy falls low enough to free neutrinos,
creating the neutrino cosmic background.

10 seconds
Electron annihilation.

Electrons and positrons annihilate,
leaving a tiny fraction of electrons remaining.
At this point the total number
of electrons equals the total number of protons.

10 seconds to 57 thousand years
Radiation era.

Photons created from
the annihilation of matter and anti matter
dominate the energy density of Universe

1- 5 minutes
Nucleosynthesis

Fusion of protons create helium,
deuterium and trace amounts of Lithium.

57 thousand years
Matter/radiation equality.

The radiation density
(photon and neutrino)
and matter density
(dark and atomic)
are equal.
This is because radiation density
falls more quickly due to the stretching
of the relativistic particles wavelengths.

Dark matter clumps into structures.
Atomic matter begins oscillation
due to the battle between gravity
and photon pressure
generating acoustic oscillations.

The first sounds of the new Universe

380 thousand years
Recombination.

The temperature falls low enough
to allow atoms to form; photons decouple.
The CMB is born, locking in its structure.

The story of the earliest times in the Universe.

5 to 200 million years
The dark age.

Photons fall into the infra red energy range;
the Universe goes dark.

The atomic gas continues to fall toward
the dark matter clumps
which grow more pronounced.

Near 100 Million years
the densest clumps
halt their expansion and begin collapsing.

By 200 Million years
the first mini halos
form and within these
the atomic cloud cools and collapses
to make the very first stars
whose light brings to an end the dark era.

200 million years
First stars.

The first stars are very massive and short lived.
They die in violent Super Nova explosions
filling the cosmos with the building
blocks of planets and the elements
needed for life.

200 to 800 million years
Epoch of ionization.

The radiation from the stars
and possibly the first quasars,
ionizes much of the remaining
neutral hydrogen and helium.

A thin mist returns
and partly obscures the CMB.
(Future Low Frequency Radio Telescopes
may soon be able to see the epoch of ionization)

1 to 2 billion years
Infant galaxies.

Star groups merge,
forming the very first galaxies.

There are frequent collisions of galaxies,
high star birth rates
and high supernova rates.

Heavy element production
changes the pattern of star formation,
making them lower mass,
less luminous and longer lived,
like those of today.

The stage is set for the emergence of life;
the Cosmos will soon
have eyes to see and minds to think.

2 to 3 billion years
Star birth and quasar peak.

In the dense environment
of frequent galaxy collisions
the star birth rate
reaches it maximum,
as does the forming and feeding
of supermassive black holes.

6 billion years
First rich galaxy clusters.

Enough time has elapsed for the densest
regions to stop expanding and form clusters.

7 billion years
Deceleration /acceleration.

The effects of Dark energy kick in.
The Universe once again begins
to accelerate its expansion rate,
but at a much more gentle rate.

8 billion years
First modern spiral galaxies.

Although some elliptical galaxies
form in the first billion years,
classic spiral galaxies aren't seen
until about 5 Billion years ago.

9 billion years
Matter / dark energy equality.

At this time the falling density
of matter (dark and atomic)
become equal to that of dark energy.

9.1 billion years
Sun and Earth form.

The solar system forms
in the outer disk of the milky way.

The stage is set for the emergence
of humankind in the Cosmos.

13.7 billion years.
Present time.

Human civilization reaches its peak
and begins heading into decline
and eventual extinction
due to over population,
resource depletion,
and environmental destruction
which generates conflict
as Human nation states fight
for ever dwindling resources.

Hopefully Humankind is not typical
and intelligent life elsewhere
solves the problem of balancing
intelligent life needs
with available resources
by developing communitarian
economic social structures.

16 to 17 billion years
The Milky way collides
with the Andromeda galaxy.

Somewhere within this time
the Sun enters into its Red Giant phase,
vaporizing the earth.
Humankind extinct for over 4 Billion years
is not around to witness this event
though possibly a new intelligent species

which emerged after
the extinction of Humankind might be.

It will be a very sad time for them
unless their technology includes
very advanced space flight.

20 billion years
Growth of Structures cease.

Expansion due to Dark energy
empties each casual patch of the Cosmos.

The story of our Universe draws to a close.

100 billion years
What remains of the Milky way
is alone in its causal patch of the Universe.

1000 billion years
Last stars die.

The Universe is empty and dark.
However, stirring in the vacuum
of space time itself
are the ever present
vacuum fluctuations.

One small patch quite by chance
fluctuates sufficiently to create a
volume of false vacuum which cuts off
from its mother Universe
by negative pressure,
explodes into a new Universe
creating new space time and future hope
for the emergence of intelligent life in the Cosmos.

Everything starts over.

< 50 >
— In the Zone —

The highest zone is absolute happiness,
'Though even the best can slip to well-being,
And sometimes, down into the bearable zone;
Next come the anger, apathy, and death zones.

< 51 >
— The (D)anger Zone —

Once we drop into the anger zone, the
Analytical mind cuts out, giving way
To the primitive reactive mind, a
Moronic state in which even beige seems black.

< 52 >
— Thus, All Become Equal —

The simple reactive mind 'thinks' that, say,
A perceived bad tone equals insult equals
Hate equals great anger equals lash out
Equals big fight equals kill equals death.

< 53 >
— The Weak Link —

The mind is quite weak in the fighting off
Of emotions, for they have a direct
Pathway into consciousness—inhibiting
The rational, thinking part of the brain.

< 54 >
— Primitive Controllers —

Emotions usually take sole control,
Brain logic relegated to the sidelines,
Being ineffective against a mood;
It's a wonder what's really is in charge.

< 55 >
— You're History —

Reason and emotion are hard to coordinate,
Each having a separate pathway to the mind;
That perhaps is all there is to tell about the
Miseries and follies of human history.

< 56 >
— The Stain on the Brain —

Emotions are slow to react to logic,
Like molasses or slow forming crystals,
Or not at all, like rocks, blocking us.
Unless and until they change, progress halts.

< 57 >
— Molecular Events —

The way that we think and feel depends but
Upon chemicals—neurotransmitters
Like Dopamine and Serotonin that
Fluctuate; so—how meaningful are strange moods?

< 58 >
— The Creative Solution Space —

Let reactions sail on by—just observe them,
But don't act on them. This puts some distance
Between you and your conditioned response,
A space which grants a modicum of free will.

< 59 >
— Free Will and Free Won't —

When extreme thoughts arrive, uninvited, as
Most thoughts do, we veto them, saying "don't",
For while we can never will that which does
The unconscious willing, we may have "free won't".

< 60 >
— 'I' —

From its safe subjective place that's free of fear,
The 'soul', our Conscious Awareness, can witness
The strange thoughts and emotions that surface
On the mind, sent there by the subconscious brain.

< 61 >
— The Drama of the Trauma —

Conscious Awareness, which can but witness,
Is a safe haven from which to observe
The drama of our lives playing in our minds,
Granting us a sobering distance from it.

< 62 >
— The Naturalization of Heaven —

Life should be euphoric, like spring fever,
As in those rare moments of ecstasy
When one is in the zone and cannot miss;
So—let all aversive substrates be removed!

< 63 >
— We Are the Cavemen —

The higher modes of being that await
The future-chemically-enhanced
Will make today's primitive mind-states
Seem as a child's tin flute to a symphony!

< 64 >
— The New Norm —

Some people are just a little bit crazy,
With a mild mental illness, some genes awry;
We call them inept, heartless, offbeat,
Even persistent, but interesting.

< 65 >
— Genetic Birth Defects —

One has, say, a learning disability,
So—do Equal Opportunity programs
Or Diversity programs discriminate
Against the handicap of being just plain dumb?

< 66 >
— Inherent Properties? —

Why should the wetness of water result
From the mix of hydrogen/oxygen?
How can cells, blood, heart, and nerves make life?
It is just so. So does matter make mind.

< 67 >
— Mind = Brain? —

Change the brain and consciousness changes too.
Take drugs and the emotions change as well.
Damage the brain and the mind's damaged too.
Consciousness emerges only from the brain.

< 68 >
— Identity Crisis? —

The brain could be the mind, and vice-versa,
So there is no need for the mind to turn
The brain's water into wine, if there's
No wine that's separate from the water.

< 69 >
— The Looking Glass —

Is consciousness emergent from the brain,
Somehow, a fundamental phenomenon?
Could it be the brain perceiving itself,
Something we might like to call the mind's 'I'?

< 70 >
— The Ultimate Basic —

Consciousness may be fundamental,
Like mass, space, and time, and would require
No explanation—just arising:
Mind: it matters; matter: ever mind?

< 71 >
— Minding Matter? —

Consciousness is irreducible in terms
Of basic entities, so, it could be that
The intrinsic properties underlying
Physical dispositions are experiential?

< 72 >
— The Same Coin? —

Energy/Awareness could be the same
Ground of Ultimate Reality,
On which both quantum-jumps/mind-thoughts are built,
Leading to the matter of brain neurons.

< 73 >
— Inside and Outside —

Pain's not the same as the nerves that cause it,
Yet, mind, apart, couldn't conserve energy.
Perhaps, our 'info' exists in two ways:
Consciously and neurologically.

< 74 >
— 'It' Permeates Reality —

Nature's made of "occasions of experience"
Instantiated into consciousness,
Even for electrons and lower life forms,
'Though worms sense but a smudge of reality.

< 75 >
— What's Left? —

In identifying consciousness,
We often confuse what is floating in
The stream of consciousness with the water itself;
Thus, we note not the sea in which we 'see'.

< 76 >
— Being —

How is conscious reality real-ized?
What directs the "spotlight of attention"?
Who's the "silent witness" that can do none other?
Who is the knower that does the knowing?

< 77 >
— Who Am 'I' —

'I' equals 'awareness' of the mind's state,
But, who or what, then, is this 'I', observing,
Which is all 'I' can do, transcending space,
Could it be the Ground-Of-All-Being itself?

< 78 >
— Awareness Explained? —

Awareness can never be an object
Of observation because Awareness
Is the very means whereby we observe.
We can't 'see' Awareness since we are it!

< 79 >
— Objection! —

We can never really understand Awareness,
A subject, because it's not itself an object
That we can be aware of—for the only tool
We have to use on it is Awareness itself!

< 80 >

— 'It' Needs Something to be Aware of —

Since the 'I' of our Awareness, that can but
Observe the mind's content, is not what it sees,
It is the universal subject, a soul
Of unperceptive immortality.

< 81 >

— The Illusion of the Self —

In consciousness, there's no distance between
The thing observed and what is observing.
They are, perhaps, one and the same, and so
It is that we seem to have a self.

< 82 >

— 'I' Exists? —

'I' am not this body—or even this thought,
For 'I' am a part of space-time itself,
Although 'I' require a mind/brain to 'look'—
For this, indeed, produces what 'I' look at.

< 83 >

— Who or What is Looking? —

The 'soul' of Awareness might be a basic
Property of space-time that can but observe
The 'self'—the contents of the brain— that is,
The portion that's currently on display.

< 84 >

— The Golden Touch —

The Midas-magic of our consciousness,
That quantum alchemist of potential,
Creates the Real from the Possible, for,
Everything it touches turns to matter!

< 85 >
— The Q —

In the eerie quantum world, all possible
And potential realities
Exist at once, in a superposition,
Until one emerges into reality.

< 86 >
— The Dual Aspect —

Electrons as waves are all spread out, and,
As such, are nowhere, having no position,
But they have direction. As particles they
Reside someplace, but have no momentum.

< 87 >
— No Trajectory —

Without a position and momentum
At the same time, electrons have no
Objective reality at all.
They go from here to there with no in-between.

< 88 >
— Quantum Weirdness —

Sub-atomics exist everywhere,
Yet nowhere, until they are measured, and
'Seen' by consciousness, so, now we sense
They're like the selfless states of meditation?

< 89 >
— The Two-Slit Mystery Solved? —

Do photons know they are being measured,
By sending out an 'offer' wave through
Both slits at once, and receive the 'handshake'
Of acceptance through just one slit, and go there?

< 90 >
— The Multiverse? —

One photon, unmeasured by man, does go
Through both slits, showing wave interference
Because, well, it may split the world in two:
One that we stay in and one that we don't?

< 91 >
— Eerie Similarities —

Minds seem to sense in another dimension,
Collapsing possibilities into reality,
Much like subatomic nonlocal effects
That link twin-particles which are FAR apart.

< 92 >
— The Mystery Solved by Bohm? —

Does back-action of a particle on
It's pilot wave, in concert with the wave's
Guidance, create the élan vital,
The stuff that consciousness is made of?

< 93 >
— The Élan Vital? —

A particle's back-action may be zero
On its guide-wave, be-coming randomness, but,
At the mind/brain level, could the brain feed
Back to the mind's guide-wave, and vice versa?

< 94 >
— The Well of Being —

The quantum substrate's the mother of all,
Of both matter and consciousness...
The quantum mind tries out new ideas
Through scenarios of consequences.

< 95 >
— The Buck Stops Here —

Does Atlas underlie the universe,
Standing upon the back of a turtle?
Is there an eternal basic substance
Or is it turtles all the way down?

< 96 >
— Composite Entities —

A God-who-is-a-person would, like us,
Be dependent on, and exist after,
The Ground of Ultimate Reality,
And so could not, in itself, be its own cause.

< 97 >
— We Are It! —

A God-which-is-not-a-person would be
The Ground of Ultimate Reality—
Energy, Awareness, or what you like—
The source of which forms/is our consciousness.

< 98 >
— Human Nature —

God said to Adam and Eve in Eden:
"Do what you like, but don't eat the apple".
Even we know that when you tell children
Not to touch something, they certainly will!

< 99 >
— Forbidden Thoughts —

Thoughts good and bad come and go, as the brain
Looks at itself without assigning values.
Still, lucky that others can't read our minds,
'Though forbidden thoughts are normal and sane.

< 100 >

— Just Think Of It —

If you try hard not to think of something,
Then you will just think of it all the more—
So if told to avoid impure thoughts, you'll
Think of people naked beneath their clothes!

< 101 >

— A Poor Example —

He murders by flame and flood; he tortures;
He entraps; he blames us for his mistakes;
He holds grudges for our ancestors' sins;
He throws tantrums and fits—his name is God.

< 102 >

— Poor Craftsmanship —

Who's to blame for the genetic defects
That lead to social misfits, obsessors,
And other special personalities?
Did the Maker's hand shake when He made us?

< 103 >

— The Way It Is —

Only a Fool would blame His own creations
For the flaws therein, for His poor craftsmanship,
So rejoice, there's no Maker of Man—these 'flaws'
Provide for interesting character types!

< 104 >

— Hallucinations and Voices —

The founders of the religions were all
"Divinely inspired", but were each told a
Different story; now we know that their
Visions might be psychotic episodes.

< 105 >
— Reasonable Doubt —

Well, possibly, probably, the quieting
Of the brain's self-boundary ID center
During focus on mantras, hymns, or prayers
Is a neurological effect, and nothing more.

< 106 >
— Tunnel Vision —

The so-called near-death experience of
Bright lights and a peeking into Heaven
Is but a flood of opiate endorphins
Causing hallucinations and calmness.

< 107 >
— Mindless Souls —

Does Awareness, our soul, have any will
Before birth or after death without a
Brain and a mind for memory, senses,
And decisions? No, not as we know it.

< 108 >
— The Time When Life is Not —

If souls are eternal, then where were they
During the eternity before births?
Nonexistence is nonexistence, whether
It comes before or after a lifetime.

< 109 >
— Enduring, But Not Unique —

There may be no distinct and enduring
Personal identity after death.
For brainless, mindless souls share, with others,
The basic Ground of Existence: Space-Time.

< 110 >
— The Problems of Traditional Religion —

The Christian concept of reward and punishment
Handed out by an omnipotent, omniscient God,
Is derivative of the family experience—
The child and parent—a conception of our world.

< 111 >
— Geographically Arbitrary —

Most deep religious beliefs are shaped by
Little more than local social forces:
Jewish, Buddhist, Islamic, Christian, or none,
So then, how deep and spiritual are they?

< 112 >
— Our Cousins —

The chimps, with whom we share 98%
Of our DNA, through common ancestors,
Are just about as special as we are,
For they have consciousness and feelings too.

< 113 >
— Ground-Of-Determination (G-O-D) —

G-O-D is merely the quantum wave function
Of the entire universe, hardly some
Vengeful and overly emotional
Superbeing who thrives on adoration.

< 114 >
— "The All" and "The One" —

Energy is eternal, for it can
Neither be created nor destroyed,
Being made but of itself, omnipresent;
It's the Mother of all Reality.

< 115 >
— To Be —

The Universe is the ultimate free lunch,
It bubbling out of 'no-where' into 'now-here',
From the quantum foam, via strings and quarks,
The Ground of Ultimate Reality.

< 116 >
— Fatherless —

The universe bubbled out of 'nothing',
Pluses forming matter; minuses residing in forces—
All in perfect balance, self-sufficient,
Needing nothing outside of itself, zilch.

< 117 >
— I Am What I Am —

I'm the All and the One, omnipresent,
For I'm eternal and can neither be
Created nor destroyed, being my own cause
And the Ground of All—I am Energy.

< 118 >
— Timeless —

Time too has a shape in 4D space-time,
And thus, like space, could have an existence
Before the birth of the universe, so,
'Creation' didn't have to wait forever

< 119 >
— Destiny with Luck —

Evolution had no real direction,
Except what was advantageous to life,
However, since we are here, the tendency
To exist was in matter all along.

< 120 >

— The End of Science? —

We cannot see beyond the quantum realm;
The dusk of physical science arrives?
Well, who knows, for nature is efficient—
So, likely, simple answers will appear.

< 121 >

— Extrinsic Shadow, Intrinsic Light —

Physics, once more direct, is now but an
Immaterial science of math-shadows,
While mysticism, once but a fogged notion,
Now's the direct observation of the Light.

< 122 >

— The Knowing —

People can't usually ever see
Further than an order of magnitude
Beyond where they are rutted, but some can
Intuit ultimate reality.

< 123 >

— The Mystical Realm —

It said, in my dreams, "Of ever waking,
It's hard to convince you in dream-language,
As when, in wakeful reality,
To tell you of that which is beyond telling."

< 124 >

— The U-R Quale —

During meditation, one clears the mind,
And so then there is no self, just one quale,
A near nothing that has a little need to be;
Is this what-it's-like to be an electron?

< 125 >
— Meditation is Not What You Think —

Meditation shifts intention away
From controlling and acquiring
Toward acceptance and observation:
One takes-in instead of acting upon.

< 126 >
— Open the Grasping Hand —

Enlightenment's not grasped or possessed,
For acquisitive aim locks the secret out—
The same mentality that one begins with.
This is why "the secret protects itself".

< 127 >
— Setting All Else Aside —

The Spiritual refers to profound connection,
Though not through visions or ecstatic emotion,
But by the experience of connectedness that
Underlies reality, and nothing more.

< 128 >
— You Are the Absolute —

Awareness is the ultimate being,
Fundamentally connecting with G-O-D;
It cannot be known in terms of worldly
Objects—it's like, well, you have to be there!

< 129 >
— To Receive is to Be —

Meditation relieves the survival self,
Shifting attention from acting to allowing,
From emotional identification to observation,
From instrumental thinking to receptive experience.

< 130 >

— Easy and Simple —

Meditation, renunciation, and service
Are not really mysterious, just different
From the usual object-oriented approach—
Mysticism is modern and ancient, not esoteric.

< 131 >

— The Job Complete —

In serving the task, one forgets the self,
Accessing the world's connected aspects,
Going beyond self-centered states of mind—since
The survival of mankind is at sake.

< 132 >

— The Fabric of Space-Time —

Being is to doing as ground is to figure,
As subject is to object as essence is to existence,
As Awareness-Consciousness is to mind-brain,
As the ultimate simplicity is to the composite.

< 133 >

— Seeing All —

The connectedness of everything to everything—
A rudimentary perception in and of itself,
Experiential in its ultimate physical disposition—
Facilitates our consciousness of interior and exterior.

< 134 >

— Cosmic Soul —

3D objects encoded into 2D,
Project as 3D holograms when laser lit;
So add dimension to your encoded self
By shining a spiritual light through it.

< 135 >
— Consciousness Really Explained —

Consciousness mediates thoughts versus outcomes
And is distributed all over the body—
From the nerve spindles to the spine to the brain—
A way to actionize without moving.

< 136 >
— The Theory of Everything —

Awareness is all there is; look no further;
Its swirling energy gives a presence
And a knowing to waves and particles
That leads, in total, to our consciousness.

< 137 >
— Everything is Possible —

Are we encodings of what happens in
The next lower dimension—in the deep
Interior of the unseen bubble where
Everything connects to/is everything?

< 138 >
— The Fifth Dimension —

Are we the encodings of holograms
Which really exist in a lower dimension,
Where everything relates to everything—
That non locale where Consciousness resides?

< 139 >
— The All Is In the One —

All things are infinitely connected,
As in a hologram—each part contains the whole;
Everything interpenetrates everything;
The universe is a seamless web of information.

< 140 >
— Quantum Memory —

In the brain's memory, every piece of info
Is cross-correlated with every other piece,
Which allows instant access and association,
For memories are stored holographically.

< 141 >
— The Prison —

I can never share a mind directly,
For there is no access; we are alone.
Mind melding works only for the Vulcans.
This loneliness leads us to company.

< 142 >
— Loneliness —

The unbearable solitude of consciousness
Is relieved by literature, social clubs,
Movies, caring, friendships, discussion, writing,
And other sharing acts, but, mostly, by love.

< 143 >
— The Devil is Dying —

Six hundred ago, the church thought that ills
Of a physical nature were caused by
Evil spirits; however, now we know
They're from bacteria and viruses.

< 144 >
— The Devil is Dead —

Now the church thinks that ills, or sins, of a
Mental nature are caused by the Devil,
An evil spirit; however, now
We know of brain chemistry gone astray.

< 145 >

— Role Model? —

Is God a good role model, a leader,
Someone that we would follow, imitate,
Emulate, be like, adore, or follow?
Well, then, what would his example provide?

< 146 >

— Not To Follow —

We could jail people for the sins of their
Ancestors, exterminate humanity,
Allow known evil to exist and tempt,
And devise devious entrapment plans.

< 147 >

— Bad Role Model —

We could have temper tantrums and outbursts,
Envy, or not permit competitors,
Grant free will only it matched our own,
And covet worship, adoration, and praise.

< 148 >

— Zero-Sum Game? —

Plus and minus from nothing came to be;
But while most charges rejoined, some went free,
The pluses forming matter, energy,
And the minuses forming gravity.

< 149 >

— Present Time and Past Space —

Mind and matter are made of the same stuff,
That substance made only out of itself.
Mind experiences the present moment;
Matter records the present from the mind.

< 150 >

— The Minding of Time, the Matter of Space —

That is, Present Mind, Past Matter, combine
The frames of Space and Time into the film
That lives and plays in us as Consciousness,
Mind taking Space and Matter doing Time.

< 151 >

— Absolute and Fundamental —

Yes, mind/matter stems from the Eternal
Substance/Space-Time/Experiential Being.
For life's great riddle of Oneness is that
Mind really Matters; Matter ever Minds.

< 152 >

— Criminal and Evil —

The right to act ends where others begin;
There is no excuse—zilch—for tactlessness.
For, if done, it causes damage, surely
Bringing down yourself and the others, too.

< 153 >

— The Sound of Silence —

When a tree falls in the forest and there's
No one around to hear it, does it make
A sound? No, for there is no ear to turn
The sound waves into sound.

< 154 >

— Nothing Seen —

Nor is there a smell, for there is no nose
For the odorous molecules to attach to.
Nor has it any color, for there is
No retina to decode the light frequencies.

< 155 >
— No Looks —

What does it look like, then? It doesn't look
Like anything, for there is no brain to
Put it all together by detecting
Form, color, texture, size, taste, smell or vision.

< 156 >
— The Derivation of a Secret —

Since the entropy of a black hole is known
To depend on the surface area of the
Event horizon and NOT on its volume,
Then our third dimension MUST BE a projection.

< 157 >
— Using the Illusion —

A projected illusion, as in a hologram,
May still be used as it were really there
Since we can make sense of it, so to speak,
But in truth the third dimension does not exist.

< 158 >
— The Bell Paradox Resolved —

Thus, apparently separate particles,
Like created photon pairs, copy the other,
When one is changed, because, in truth, they are
Still the same thing in the projector room.

< 159 >
— What the Tree 'Looks' Like —

If the universe is holographic,
Then the tree in the forest, whether seen or not,
Is, at heart, an interference pattern
Brought to life only when we tune it in.

< 160 >

— Our Model of Reality —

This is the mystery of the realness
Of sleeping dreams revealed: we tune in to
The interference patterns, whether awake
Or asleep, to bring alive the reality projected.

< 161 >

— All is One —

Everything connects to everything else
Through overlapping interference patterns,
And so nothing is separate at all, as it seems,
But is one large all-encompassing whole.

< 162 >

— Instant Recall —

Memory, too, seems to be holographic,
Residing everywhere in the brain,
Every piece associated with others related,
Instantly broadcasting all the connections.

< 163 >

— Blake's Vision Confirmed —

Every part of a hologram contains the whole,
The whole universe contained within a
Grain of sand, all eternity within a moment,
The universe rumbling when an electron vibrates.

< 164 >

— Deepak's Ultimate —

We are part and parcel of everything—
We are the cosmos; we are life; we are love;
We are all that is; we are the creator
Of the dance as well as the dancer.

< 165 >
— To Be Investigated —

Whether the past is recorded and accessible
As part of the holographic whole is not known
Or whether the other two dimensions are
Projected, as well, but perhaps we shall see.

< 166 >
— The Eternal Substance —

This then is the secret of the universe,
Knowing of that which underlies all reality:
Fundamental, absolute, indestructible,
Omnipresent, omnipotent, and all pervasive.

< 167 >
— IT is What IT Is —

Why absolute and fundamental? Because
It is made of one piece—itself,
And therefore indestructible, and eternal, too,
And makes up all that there is, everywhere.

< 168 >
— M = E/CC —

Perhaps matter is more than just equivalent
To energy if it were transformed,
And more than equal to it, more or less—
It may be that matter IS energy.

< 169 >
— E = ? —

Perhaps energy is projected by
Information or is a part of our
Reality illusion, as well, but,
We can't stand on turtles all the way down!

< 170 >
— Little Matter —

Only four percent of the Universe
Consists of the matter we know and love,
The remainder being dark and hidden,
So perhaps this is what's really in charge.

< 171 >
— Nothing is Impossible —

The basis of the Universe was forever here,
For nothing can make itself from nothing at all;
Such, a state of nothing could never be, for there IS
Something—something that consciousness interprets.

< 172 >
— Through the Haze —

The Infinite may radiate through a matrix,
Using Information or Energy to create
The Cosmic Background antenna which broadcasts
Interference patterns of virtual reality.

< 173 >
— Creating Life —

Well, how can I save the Soul—Consciousness?
It may create potentials, quantum-like,
That give rise to the Reality of
The Mind and Body, so use it wisely.

< 174 >
— Defeating Natural Determinism —

Choose what is good for everyone, not just you,
After careful philosophic thought,
Then do it, even though your natural
Inclination may be not to do it.

< 175 >
— The Third Millennium —

The secrets of the universe are all found—
All exists out of consciousness, the ground.
Blame, soul, free will, and God all have fallen—
But it will take thousands of years to sink in!

< 176 >
— Beyond Local Reality —

Time, space, stuff, change, and form were real-ized from
The Fundamental Possibility,
Becoming our penultimate reality—
One possible from all probabilities.

< 177 >
— Quantum Superposition is Real —

Our reality came not from nothing,
But existed always as possibility,
One that amounts to something workable,
Among all in superposition.

< 178 >
— The First Impossibility —

No form of our penultimate realness
Could have existed alone before
Everything was quantum-known-all-at-once,
For what could have made the choice among many?

< 179 >
— The Second Impossibility —

Nor came it from an absolute nothing,
Since there can be no such ‘thing’ at all,
So, since either way is impossible,
Fundamental Possibility is.

< 180 >
— The Unbelievable Truth —

This ultimate basis of reality
Though not much like our local reality,
Is hinted at by quantum physics—
It forms reality real as can be!

< 181 >
— The Verifiable Truth —

So how else could it be, for particles
Do appear and disappear from somewhere,
Going from here to there with no between,
Manifesting from no-where to now-here.

< 182 >
— The Search —

I'll follow every single avenue,
Whether it's brightly lit or a dark alley,
Exploring one-ways, no-ways, and dead-ends
Until cornered where the truth is hiding.

< 183 >
— The Question —

Since we all became of this universe,
Should we not ask who we are, whence we came?
Insight clefts night's skirt with its radiance—
The Theory of Everything shines through!

< 184 >
— The Sum —

Some simple substances gave rise to everything,
Chosen as probable above the rest—
Known all-at-once that it would be the best—
The most promising—the possible ones.

< 185 >
— The Scale —

As to how complex, there is no limit,
But to collapse into a black hole;
The smallest of all is the planck distance,
So size is absolute, not relative.

< 186 >
— Live —

Like the moon, challenge night and gain the light;
Like the rose, suffer the thorn—gain the fragrance;
Of life, surrender to live forever—
Enlightened more than a thousand suns.

< 187 >
— Be —

World does not pass by—you pass through it;
Clear your being so the treasure may arrive;
This spirit sparkles of a different light—
The gemstones are of a different mine.

< 188 >
— Unrealized Power —

Mind reaches out to see what's possible
And what's not, like particles forming
In the quantum world, but, better than that—
It makes the impossible possible.

< 189 >
— Mind/Consciousness is Heaven —

Mind is the ultimate of all there is
It is the universe—billions of years
Of primordial material—complex.
So, what more could human beings want?

< 190 >
— Electromagnetic Unity —

Electricity and magnetism each
Lead to the other, being transformational.
They facilitate action and motion
Through EM's push-pull of regularity.

< 191 >
— Strong/Weak Opposition —

The strong force binds the atomic nucleus
Barely beating EM's repelling force.
The weak force counters strong's stability
Through decay that promotes changeability.

< 192 >
— Electroweak Unification —

Electromagnetism and the weak force
Unify when the temperature gets hot,
As during the Big Bang, but they oppose
The strong force as duality's balance.

< 193 >
— AWOL —

What about gravity? Where has it been?
It needs matter and motion to exist
And so it is the blended result of
All the forces, a secondary effect.

< 194 >
— The Duos and Duels of Nature —

Dualities seem to assist nature:
Good/evil, on/off, hot-cold, man/woman,
Up/down, left-right, here-there, past-future, and
So, none can exist without the other.

< 195 >
— Separation Allows Duality —

There can be no more unification,
For what One could be versatile enough
To form both the electroweak force and the
Strong—as different as the north/south poles.

< 196 >
— A Smooth Transition —

Past that was leads to future that will be—
Transformational—'now' in the middle
Rolling smoothly through recall, sensation,
And anticipation. Time is movement!

< 197 >
— Space/Matter Duality —

Matter forms inertial knots in space's place,
While space places and separates the knots.
Open-endedness counters form's closure
In the ying-yang cycle of appearance.

< 198 >
— The Élan Vitale —

Space/matter and past-future blend to create
The spirit of life as the pyramid's core
That furthers the sparks of pair relationships
That evolve as the life of our species.

< 199 >
— What Matter Was and Will Be —

Past matter is history—what's occurred,
While future matter is progression seen.
Matter past to future changes structure,
That which moves and/or reforms through time.

< 200 >
— What Space Was and Will Be —

Past space is remembrance—the memories,
While future space is wishes, hopes, and dreams.
Space past to future is a change of outlook—
From what is known to what might become.

< 201 >
— Past & Future & Space & Matter —

Remembering history is learning;
Wishing of a progression is vision;
We venture on into creation from
Structural changes and education;

< 202 >
— Continued —

Direction is learning from outlook's change,
While planning's the formation of vision;
Vision and change of outlook beget growth;
All of no-where to now-here as Being.

< 203 >
— Not-Here to Every-Where —

Life, mind, spirit, form, time, and consciousness
Derive from the fundamental content
That materialized from the unknowable,
And grant us the experience of being.

• • •

VIOLATING UNIVERSAL NATURAL LAW

You will always be caught,
So don't even give it a thought.

The violation of universal natural law
Is the cause of our problems, all,
Of everything that becomes rife
And plagues individual and national life,
These stresses only leading to more strife,
From lowlifes leaving their wife for the wildlife
Of nightlife to cutting someone with a knife.

So stem problems of national health,
Crime, the economy, education, wealth,
And the black environmental sins,
All of them having their origin
In a widespread law violation
By some portion of the population.

Universal Natural Law is very terse
In governing the entire universe,
It being the orderly principles
That regulate physical events/processes.

Science defines the universal law of nature,
A precise description of how nature nurtures.

Universal law pervades everything,
Of all that is in passage and being,
From the motion of particles
To the evolution of life's articles—
Operating at every scale:
The subatomic, atomic,
Molecular, biological, geological,
Astrophysical, and cosmological.

The universe is structured hence
In these many layers of existence
As worlds within worlds,
Distinguished and not only furled
By vastly different time and distance scales,

But that every level has its own set of details;
For example, an electron/nucleus system
Is not analogous to that of a planet/sun.

The more superficial macroscopic levels of nature
Can be seen as fragmented expressions, for sure,
That are manifested from the more unified laws
Governing deeper levels with their scrimshaws—
The reflections of the dazzling symmetries
Of what once were inaccessible mysteries.

The outer 'becomes' are based on inner ones,
The only fountainhead of all the rhythms.
(And the converse is not true.)

Nature's governance is maximally efficient,
For it is frugal, and not a spendthrift—
It following The Principle of Least Action
In all of its action and protraction.

This is why a ray of light refracts
When going from air to water's tract,
Minimizing the time
And saving every dime.

From this maximal economy of nature,
All classical behavior can be scriptured.

Entropy is a count of quantum states
Accessible to a macroscopic system's estate,
This available number ever increasing;
The nature of life is to grow, ever reaching.

The path of least action's welcome
Is just the macroscopic outcome
Of the simultaneous superposition
Of multiple coexisting paths' auctions
At the microscopic level,
The outcome ever of the least income.
The law to which all must succumb.

All is rooted in the verse
Of the Constitution of the Universe.

Life takes advantage and cause
Of the universal natural laws,
Even such as in merely walking,
Which is an immensely complex undertaking.

We employ technology
In all of its variety.

Everything that we fail to accomplish
Is but due to the total failure
To apply universal natural law effectively,
This being the source of all difficulty.

In the absence of knowledge of a lever,
The simple task of moving a boulder
Becomes complex and arduous to the shoulder.

Not learning gravity has caused unmild
Injuries to many a young child;
The old uses of radiation caused cancer tumults;
The use of DDT had many adverse results.

Smoking cigarettes, heavy drinking, being out late,
And other addictive obsessions surely violate
Universal natural law, at whatever rate,
Resulting in negative consequences,
While psychological violations dispense
Stress directly a a sequence immense.

While fulfillment of desire can bring happiness,
It also raises the scope and standardness
Of future desires, making the duress
Of frustration an inevitable process.

Over time this causes psychological stress,
Which in turn impairs creativity's success,
Stalling future desires
By watering their fires

And also leads to problems of health,
These then causing further stealth
And violations of universal natural law—
Resulting in the nonsense
Of a life out of balance—
Leading to aggression, anxiety,
Impulsive violet behavior, hostility
And substance abuse—
A vicious cycle of refuse
That, among other effects,
Fills up the prisons to correct.

X-X vs. X-Y CHROMOSOMES

So, I see and like that female
Are doubly rated X,
And when women do strange things,
Men must scratch their heads and wonder 'Y'.

SIMPLETONS ANONYMOUS

Occam was running
An anonymous support group
Called *On and on anon.*

His opening remarks took but a few seconds:
"One can often say more with less."

Bill Clinton, along with many other politicians
And lawyers, then went on and on for hours.

WILLING THE WILL THAT WILLS?

[Revelations: Austi: 4:5]

What is the "secret" of human behavior,
One that's really so much the saviour
That we may even keep it from ourselves
Rather than very far into it try to delve?

What is it that should be so confidential,
Classified, and undisclosed—its potential
Kept under wraps, so very contra;
Informally: hush-hush; formally: sub rosa?

Well, it's a revelation of splendor,
One that's often good to surrender
But is also very well to remember.

Is the will free to will one's actions otherwise?

Can antecedent conditions be ignored?

Can the self be an unmoved mover?

No, but...

And what of those tendencies of evo's realm
That have been imprinted on one's genetic film—
Those of temperament, role preferences,
Emotions, responses,
And even one's most revered moral choices—
Those invoices from which one rejoices?

Well, these are not choices
At all in of any free will voices;

In essence, from the basis of one
And from all that one has become
From life's total behavioral reactions,
There are probabilities of actions—
Some patterns that are very likely
And some patterns highly unlikely.

Is free will a necessary fiction,
A kind of a religion?

No, and yes if it's to provide an essential berth
For one's morality, meaning, and worth.

So, then, with this "free will" become,
One might then succumb
To systematic deception
About one's causal connection
To that of nature,
A roadblock, a detour
That's neither possible,
Necessary, nor desirable.

The enemies to these "free will" motifs
Would be the mythical cultural beliefs
That explain behaviors and feelings
In terms of unknowable forces and beings.

But, to protect one's moral virtues
Should one still believe oneself's purview
To be as an ultimately responsible agent, lo—
A self creation exnihilo,
A God-like, miniature first cause who chooses
Without it being determined by one's own muses?

Well, maybe, but, nay, really not, nil,
For there is no contra-causal free will.

What the good then
Of this fix we're in?

Such it is then that we can gain a measure of peace
Rather than the anger of resentment's crease
When someone does or says something bad,
Even those close relatives you once had.

For the civil-law-breakers
And all those ungiving takers
We'll no longer incarcerate
For punishment, being so irate at the jailbait,

But so that society will be protected
And that they might emerge corrected
From the swill of a prison mill,
Fulfilled with a new unfree will
That points more toward goodness
Or at least away from badness.

Thus, the action
Of metaphysical justification
For a total retribution
Then greatly softens,
A relief from the stress, so often,
For it's no longer induced
From the abuse produced.

Really?

Truly.

Indeed, we become less self-conscious,
More playful, less noxious, more gracious,
Less callow, and less likely to wallow
In the sorrow that is so hollow and shallow
In its excessive self-blame, pride,
Envy, or resentment—now all put aside.

Aren't we changing the will here as we go?

Yes but mostly no,
For the will must ever follow what we know.

Then we are learning—
The only hope for larger earnings
From the will's then wider yearnings!

Yes, overturning.

What if to learning we are averse?

What a curse! Might as well call the hearse.

So, then, all in all, though a tempt,

It is that we humans are not exempt
From the laws of physics—a preempt
Although we've been wired to make the attempt—
A seeming violation by nature
Of its own universal law and structure.

No, it's not a violation I would call,
For science still did tell us all.
It's all part of the structure;
One can never cheat Mother Nature.

Hail, then, to the physic.

Well, it's not so bad, is it?—
Although we can never will the will,
Its motives ever our intent to fulfill;
It is that we have no free will.

True, plus we can expand the will's horizoning
Through our broader learning's wisening.

Yes, learn today and by tomorrow, say,
The will may have a different sway.

I wouldn't want it any other way,
For then I wouldn't be me—my screenplay.

What other ways can we improve the play?

Well, we have patience and delay,
For we don't have to act right away.

Until a more creative solution appears?

Yes, from any frontier, Shakespeare.

Hear, hear!

LIFE'S FIRST WORDS

The Earth is ~ 4.5 billion years old.
When it was ~ 3.8 billion years old
And capable of supporting life,
LIFE, in the form of a single cell entity,
Came to be.
No evolution here as there
Was nothing to evolve from,
And on an evolutionary scale
It was almost instantaneous.
Rather than evolution it was more like emergence.
A very, very complex form of emergence.
What force of nature can explain this?

It took ~ 3.8 billion years for man
To evolve from that single cell.
That to me makes sense, that is evolution.

Now I believe there is an answer to this,
Because this is what happened,
But what is the answer???

And the first life form said,
"Where the heck is everyone?"

But that saying is long forgotten,
Since it then said
"This is one small step for a bug,
But a giant leap for mankind."

(In Australia, it was said
"A giant leap for a kangaroo".)

— THE END —

www.ingramcontent.com/pod-product-compliance
Lightning Source LLC
Chambersburg PA
CBHW071358170526
45165CB00001B/99

* 9 7 8 1 4 5 2 8 2 2 6 8 6 *